Matthew E. Taylor

Transfer in Reinforcement Learning Domains

Studies in Computational Intelligence, Volume 216

Editor-in-Chief
Prof. Janusz Kacprzyk
Systems Research Institute
Polish Academy of Sciences
ul. Newelska 6
01-447 Warsaw
Poland
E-mail: kacprzyk@ibspan.waw.pl

Further volumes of this series can be found on our homepage: springer.com

Vol. 195. Vivek Bannore and Leszek Swierkowski
Iterative-Interpolation Super-Resolution Image Reconstruction:
A Computationally Efficient Technique, 2009
ISBN 978-3-642-00384-4

Vol. 196. Valentina Emilia Balas, János Fodor and Annamária R. Várkonyi-Kóczy (Eds.)
Soft Computing Based Modeling in Intelligent Systems, 2009
ISBN 978-3-642-00447-6

Vol. 197. Mauro Birattari
Tuning Metaheuristics, 2009
ISBN 978-3-642-00482-7

Vol. 198. Efrén Mezura-Montes (Ed.)
Constraint-Handling in Evolutionary Optimization, 2009
ISBN 978-3-642-00618-0

Vol. 199. Kazumi Nakamatsu, Gloria Phillips-Wren, Lakhmi C. Jain, and Robert J. Howlett (Eds.)
New Advances in Intelligent Decision Technologies, 2009
ISBN 978-3-642-00908-2

Vol. 200. Dimitri Plemenos and Georgios Miaoulis *Visual Complexity and Intelligent Computer Graphics Techniques Enhancements,* 2009
ISBN 978-3-642-01258-7

Vol. 201. Aboul-Ella Hassanien, Ajith Abraham, Athanasios V. Vasilakos, and Witold Pedrycz (Eds.)
Foundations of Computational Intelligence Volume 1, 2009
ISBN 978-3-642-01081-1

Vol. 202. Aboul-Ella Hassanien, Ajith Abraham, and Francisco Herrera (Eds.)
Foundations of Computational Intelligence Volume 2, 2009
ISBN 978-3-642-01532-8

Vol. 203. Ajith Abraham, Aboul-Ella Hassanien, Patrick Siarry, and Andries Engelbrecht (Eds.)
Foundations of Computational Intelligence Volume 3, 2009
ISBN 978-3-642-01084-2

Vol. 204. Ajith Abraham, Aboul-Ella Hassanien, and André Ponce de Leon F. de Carvalho (Eds.)
Foundations of Computational Intelligence Volume 4, 2009
ISBN 978-3-642-01087-3

Vol. 205. Ajith Abraham, Aboul-Ella Hassanien, and Václav Snášel (Eds.)
Foundations of Computational Intelligence Volume 5, 2009
ISBN 978-3-642-01535-9

Vol. 206. Ajith Abraham, Aboul-Ella Hassanien, André Ponce de Leon F. de Carvalho, and Václav Snášel (Eds.)
Foundations of Computational Intelligence Volume 6, 2009
ISBN 978-3-642-01090-3

Vol. 207. Santo Fortunato, Giuseppe Mangioni, Ronaldo Menezes, and Vincenzo Nicosia (Eds.)
Complex Networks, 2009
ISBN 978-3-642-01205-1

Vol. 208. Roger Lee, Gongzu Hu, and Huaikou Miao (Eds.)
Computer and Information Science 2009, 2009
ISBN 978-3-642-01208-2

Vol. 209. Roger Lee and Naohiro Ishii (Eds.)
Software Engineering, Artificial Intelligence, Networking and Parallel/Distributed Computing, 2009
ISBN 978-3-642-01202-0

Vol. 210. Andrew Lewis, Sanaz Mostaghim, and Marcus Randall (Eds.)
Biologically-Inspired Optimisation Methods, 2009
ISBN 978-3-642-01261-7

Vol. 211. Godfrey C. Onwubolu (Ed.)
Hybrid Self-Organizing Modeling Systems, 2009
ISBN 978-3-642-01529-8

Vol. 212. Viktor M. Kureychik, Sergey P. Malyukov, Vladimir V. Kureychik, and Alexander S. Malyoukov
Genetic Algorithms for Applied CAD Problems, 2009
ISBN 978-3-540-85280-3

Vol. 213. Stefano Cagnoni (Ed.)
Evolutionary Image Analysis and Signal Processing, 2009
ISBN 978-3-642-01635-6

Vol. 214. Been-Chian Chien and Tzung-Pei Hong (Eds.)
Opportunities and Challenges for Next-Generation Applied Intelligence, 2009
ISBN 978-3-540-92813-3

Vol. 215. Habib M. Ammari
Opportunities and Challenges of Connected k-Covered Wireless Sensor Networks, 2009
ISBN 978-3-642-01876-3

Vol. 216. Matthew E. Taylor
Transfer in Reinforcement Learning Domains, 2009
ISBN 978-3-642-01881-7

Matthew E. Taylor

Transfer in Reinforcement Learning Domains

Matthew E. Taylor
Postdoctoral Research Associate
Department of Computer Science
The University of Southern California
Los Angeles, CA 90089
USA
E-mail: taylorm@usc.edu

ISBN 978-3-642-10186-1 e-ISBN 978-3-642-01882-4

DOI 10.1007/978-3-642-01882-4

Studies in Computational Intelligence ISSN 1860949X

Typeset & Cover Design: Scientific Publishing Services Pvt. Ltd., Chennai, India.

Printed in acid-free paper

9 8 7 6 5 4 3 2 1

springer.com

This monograph is dedicated to my parents.

Foreword

It is a great pleasure and honor to be able to write the forward for this book, representing the culmination of Matthew Taylor's Ph.D. thesis research at The University of Texas at Austin. It was my good fortune to be Matthew's advisor during his five years as a Computer Sciences graduate student. I therefore was able to participate in and enjoy the adventure of starting with the kernel of an idea and fully developing it into a full-blown dissertation. As is always the case, there were pitfalls along the way. But in all cases they were quickly overcome, and often turned into strengths of the research.

As is fully reflected in this book, Matthew's dissertation research centers around the challenge of enabling a learning agent to improve or speed up its learning on one task based on previous experience with a similar, but different task. This *transfer learning* challenge is a current important topic in the field of Artificial Intelligence. The distinguishing feature of Matthew's research is that he focuses on reinforcement learning tasks (tasks characterized by sequential decision making and delayed reward) and has shown that it is possible to successfully transfer knowledge encapsulated in the value function (an internal representation of the expected reward from each state). He has demonstrated that it is possible to transfer knowledge between tasks in very different domains, and to automatically learn the inter-task mapping that matches the state variables and actions between the tasks.

This research is very important for the field because, in the past, most machine learning systems have focused on training and testing on the same problem. For example, a checkers-playing program can improve its performance by playing many checkers games. But often, it is necessary for an agent to be able to leverage past experiences in different tasks towards new ones. Ideally an agent that has learned to play checkers should be better at learning chess than an agent learning chess from scratch. This transfer ability is characteristic of humans. But prior to Matthew and a small group of our colleagues, it was not possible for computer programs.

By way of the research presented in this book, Matthew has established himself as the pre-eminent worldwide expert on transfer learning in sequential decision making tasks. A particular strength of the research is its very thorough and methodical empirical evaluation, which Matthew presents, motivates, and analyzes clearly in prose throughout the book. Whether this is your initial introduction to the concept of transfer learning, or whether you are a practitioner in the field looking for nuanced details, I trust that you will find this book to be an enjoyable and enlightening read.

Austin, TX
February 14, 2009

Peter Stone
Associate Professor of Computer Science

Contents

Chapter 1
Introduction

In *reinforcement learning* [Sutton and Barto (1998)] (RL) problems, learning agents execute sequential actions with the goal of maximizing a reward signal, which may be time-delayed. For example, an agent could learn to play a game by being told whether it wins or loses, without ever being told the "correct" action. The RL framework has gained popularity with the development of algorithms capable of mastering increasingly complex problems. However, when RL agents begin learning *tabula rasa*, mastering difficult tasks is often slow or infeasible, and thus a significant amount of current research in RL focuses on improving the speed of learning by exploiting domain expertise with varying amounts of human-provided knowledge. Common approaches include deconstructing the task into a hierarchy of subtasks (c.f., MAXQ [Dietterich (2000)]), finding ways to learn over temporally abstract actions (e.g., using the options framework [Sutton et al. (1999)]) rather than simple one-step actions, and abstracting over the state space (e.g., via function approximation [Sutton and Barto (1998)]) so agents may efficiently generalize experience.

This monograph examines one such general method for speeding up learning: *transfer learning* (TL). The key insight behind transfer learning is that generalization may occur not only within tasks, but also *across tasks*. This insight is not new; transfer has been studied in the psychological literature for many years [Thorndike and Woodworth (1901), Skinner (1953)]. More relevant are a number of approaches that transfer between machine learning tasks [Caruana (1995), Thrun (1996)], planning tasks [Fern et al. (2004), Ilghami et al. (2005)], and within cognitive architectures [Laird et al. (1986), Choi et al. (2007)]. However, TL for RL tasks has only recently been gaining attention in the artificial intelligence community.

Transfer learning in RL is an important topic to address at this time for three reasons. First, in recent years RL techniques have achieved notable successes in difficult tasks which other machine learning techniques are either unable or ill-equipped to address (e.g., backgammon [Tesauro (1994)], job shop scheduling [Zhang and Dietterich (1995)], elevator control [Crites and Barto (1996)], helicopter control [Ng et al. (2004)], Robot Soccer Keepaway [Stone et al. (2005)], and Server Job Scheduling [Whiteson and Stone (2006)]) and quadruped locomotion

M.E. Taylor: Transfer in Reinforcement Learning Domains, SCI 216, pp. 1–13.
springerlink.com

[Saggar et al. (2007), Kolter et al. (2008)]). Second, classical machine learning techniques such as rule induction and classification are sufficiently mature that they may now easily be leveraged to assist with TL. Third, promising initial results show that not only are such transfer methods possible, but they can be very effective at speeding up learning. The 2005 DARPA Transfer Learning program [DARPA (2005)] helped to significantly increase interest in transfer learning. There have also been some recent workshops providing exposure for RL techniques that use transfer. The 2005 NIPS workshop, "Inductive Transfer: 10 Years Later," [Silver et al. (2005)] had few RL-related transfer papers, the 2006 ICML workshop, "Structural Knowledge Transfer for Machine Learning," [Banerjee et al. (2006)] had many, and the 2008 AAAI workshop, "Transfer Learning for Complex Tasks," [Taylor et al. (2008a)] focuses on RL.

With motivations similar to those of *case based reasoning* [Aamodt and Plaza (1994)], where a symbolic learner constructs partial solutions to the current task from past solutions, the primary goal of transfer learning is to autonomously determine how a current task is related to a previously mastered task and then to automatically use past experience to learn the novel task faster. This monograph focuses on the following question:

> Given a pair of related RL tasks that have different state spaces, different available actions, and/or different representative state variables,
>
> 1. how and to what extent can agents transfer knowledge from the source task to learn faster or otherwise better in the target task, and
> 2. what, if any, domain knowledge must be provided to the agent to enable successful transfer?

The primary contribution of this monograph is to answer the first part of the above question by demonstrating that TL is feasible. For this purpose, we introduce *inter-task mappings*, a construct that relates pairs of tasks that have different actions and state variables. Inter-task mappings are the field's first [Taylor and Stone (2004)] construct to enable such transfer techniques and are formalized in Chapter 5.

The first TL method to use inter-task mappings is *Value Function Transfer* (Section 5.2), which can transfer between agents[1] in tasks with different state variables and actions, assuming that both agents use *temporal difference* (TD) learning algorithms and represent the learned value function in the same manner. Experiments demonstrate that this method can significantly improve learning, but there may be situations where it is inapplicable because an agent does not use TD learning, or

[1] It is reasonable to frame TL as transferring from an agent in a source task to a *different* agent in a target task, or to consider training an agent in a source task and then having it *move* into the target task. In this monograph we discuss our methods assuming transfer between different agents, but the two views are equivalent.

because agents use different representations. However, the inter-task mapping construct is robust enough to work in a variety of settings, and this monograph fully explores the application of inter-task mappings as the core of multiple algorithms. This monograph introduces the following methods, all of which utilize inter-task mappings:

1. Value Function Transfer (Section 5.2) is described above.
2. Q-Value Reuse (Section 6.1) builds on Value Function Transfer by directly reusing a learned source task action-value function, allowing for transfer between TD agents with different function approximators.
3. Policy Transfer (Section 6.2) modifies the structure and weights of neural network action selectors to transfer between direct policy search learning methods.
4. TIMBREL (Section 7.1) directly transfers experience data between tasks in order to improve learning on a model-based learning method in the target task, without placing any requirements on the type of source task learning method.
5. Rule Transfer (Section 7.2) learns production rules that summarize a source task policy learned with any RL method, and provides the rules as advice to a TD learner in a target task.
6. Representation Transfer (Section 7.3) allows experience from an agent trained in a task to be reused in the same task by an agent with a different representation (as defined by the learning algorithm, the function approximator, and the function approximator's parameterization), or in a different task.

As a whole, these methods show that inter-task mappings can be used as a core component in multiple algorithms, allowing for transfer between many different types of learners and learning representations. Additionally, these methods show that different types of knowledge can be successfully transferred, emphasizing that inter-task mappings are a very general construct that allow for significant flexibility in specific transfer algorithms.

A second contribution of this monograph is to answer the latter part of the above question by demonstrating that inter-task mappings can be learned autonomously. While all of the TL methods in this monograph function well with mappings provided by a human, a human may sometimes be unable to generate such a mapping, either because he does not have the requisite domain knowledge, or because the agent is fully autonomous. A pair of mapping-learning methods are therefore introduced to address this potential shortcoming. The first uses classification, in conjunction with some limited domain knowledge, to learn a mapping between two tasks. The second gathers data in both tasks, uses regression to learn a simple model, and then selects an inter-task mapping by testing different possible mappings against the model offline. This second method, as discussed in related work (Section 3.6), is significantly more robust than existing methods that learn such relationships between tasks, and is capable of enabling autonomous transfer.

As demonstrated by the variety of related work in Chapter 3, there are many ways to formulate and address the transfer learning problem. This work differs from existing approaches in three ways:

1. The TL methods enumerated above use inter-task mappings to transfer between tasks with differences in the action space and state variables, which increases their applicability (relative to many existing transfer methods). Our algorithms are also applicable when the transition function, reward function, and/or initial state differ between pairs of tasks.[2]
2. Our methods are competitive with, or are able to outperform, other transfer methods with similar goals.
3. We introduce two methods that are able to *learn* inter-task mappings in order to define relationships between pairs of tasks without relying on a human to provide them. Such methods are necessary for achieving autonomous transfer but remain a relatively unexplored area in the literature.

1.1 Problem Definition

All transfer learning algorithms use one or more *source tasks* to better learn in a *target task*, relative to learning without the benefit of the source tasks. Transfer techniques assume varying degrees of autonomy and make many different assumptions. In order for an agent to autonomously transfer, it would have to perform all of the following steps:

1. Given a target task, select an appropriate source task from which to transfer, if one exists.
2. Learn how the source task and target task are related.
3. Effectively transfer knowledge from the source task to the target task.

Although no TL methods are currently capable of robustly accomplishing all three goals, there has been research on each independently. In this monograph, six TL algorithms are introduced for step 3, two for step 2, and one demonstrates initial progress on step 1. The key commonality to all of these algorithms is the use of inter-task mappings.

A TL agent leverages experience from an earlier task to learn its current task. A common formulation for this problem is to present an agent with a pair of tasks in sequence, where the first task is implicitly the source and the second task is the target. Such a formulation, as used in this monograph, requires the agent to perform step #3 above, and possibly steps #1 or #2. An alternate formulation, not studied in this monograph, allows an agent to learn a set of source tasks, in series or in parallel, and then transfer knowledge from one or more of them to speed up learning in a target task (in the spirit of *multitask learning* [Caruana (1995)] or *lifelong learning* [Thrun (1996)]). This second formulation necessarily emphasizes step #1, as there are many tasks to potentially transfer from.

Past research on transfer between reinforcement learning tasks has focused on step # 3 above, demonstrating that knowledge from a source task can be used to

[2] This monograph uses the *Markov Decision Process* [Puterman (1994)] (MDP) framework, summarized in the following chapter with explanations of its components such as the transition function and the reward function. More specifics regarding how tasks are allowed to differ are discussed when introducing our first TL method in Chapter 5.

learn a target task faster. Existing methods consider pairs of tasks with a variety of differences. For example, the source task and target task may differ in:

1. Transition function: Effects of agents' actions differ [Selfridge et al. (1985)]
2. Reward structure: Agents have different goals and get rewarded for different behavior [Singh (1992)]
3. Initial state: Agents start in different locations over time [Asada et al. (1994)]
4. Goal state: The terminal state that provides a reward to an agent is allowed to move [Fernandez and Veloso (2006)]
5. State space: Agents act in different environments [Andre and Russell (2002)]
6. State variables: The way in which agents describe their environment differ [Torrey et al. (2005)]
7. Actions: Agents may execute different actions [Torrey et al. (2005)]

Of these differences, the final two are the most difficult because transferred knowledge must be significantly modified in order to usefully apply to the target task. When physical or virtual agents are deployed, any mechanism that allows for faster learned responses in a new task has the potential to greatly improve their efficacy. As this monograph shows, although inter-task mappings were designed to allow TL methods to handle tasks with the final two differences in the list above, our TL methods are compatible with all of them; this makes TL methods more useful in practice, because they can transfer among a larger set of tasks than if only some of the above differences could be handled.

1.2 Evaluating Transfer Learning Methods

Transfer techniques assume varying degrees of autonomy and make many different assumptions. To be fully autonomous, an RL transfer agent would have to perform all of the following steps:

1. Given a target task, select an appropriate source task or set of tasks from which to transfer.
2. Learn how the source task(s) and target task are related.
3. Effectively transfer knowledge from the source task(s) to the target task.

While the mechanisms used for these steps will necessarily be interdependent, TL research has focused on each independently, and no TL methods are currently capable of robustly accomplishing all three goals.

A key challenge in TL research is to define evaluation metrics, precisely because there are many possible measurement options and algorithms may focus on any of the three steps above. For instance, it is not always clear how to treat learning in the source task: whether to charge it to the TL algorithm or to consider it as a "sunk cost." On the one hand, a possible goal of transfer is to reduce the overall time required to learn a complex task. In this scenario, a *total time scenario*, which explicitly includes the time needed to learn the source task or tasks, would be most appropriate. On the other hand, a second reasonable goal of transfer is to effectively

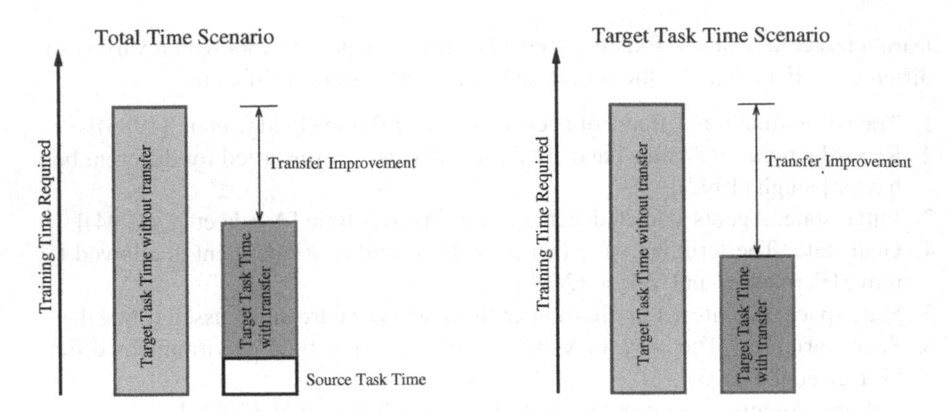

Fig. 1.1 Successful TL methods may be able to reduce the total training time (left). In some scenarios, it is more appropriate to treat the source task time as a sunk cost and test whether the method can effectively reuse past knowledge to reduce the target task time (right).

reuse past knowledge in a novel task. In this case, a *target task time scenario*, which only accounts for the time spent learning in the target task, is reasonable.

The total time scenario may be more appropriate when the agent is guided by a human. Suppose that a researcher wants an agent to learn how to perform a task, but recognizes that it may be easier if the agent learns a sequence of tasks instead of tackling the difficult task directly. The human can construct a series of tasks that the agent can train on, and possibly suggest to the agent how the tasks are related. Thus the agent's TL method will have to do little or no work for steps 1 and 2 above, but does need to efficiently transfer knowledge between tasks (step 3). To succeed, the agent would have to learn the entire sequence of tasks faster than if it had spent its time learning the final target task directly (see the total time scenario in Figure 1.1).

The target task time scenario is more appropriate for a fully autonomous learner. A fully autonomous agent must be able to perform steps 1–3 on its own. However, metrics for this scenario do not need to take into account the cost of learning source tasks. The target task time scenario emphasizes the agent's ability to use knowledge from one or more previously learned source tasks without being charged for the time spent learning them (see the target task time scenario in Figure 1.1). In this monograph we will see that the majority of existing transfer algorithms assume a human-guided scenario, but disregard time spent training in the source task. When discussing individual TL methods, we will specifically call attention to the methods that do account for the total training time and do not treat the time spent learning a source task as a sunk cost.

Many metrics to measure the benefits of transfer are possible (shown in Figure 1.2, replicated from our past transfer learning work [Taylor and Stone (2007a)]):

1. *Jumpstart*: The initial performance of an agent in a target task may be improved by transfer from a source task.
2. *Asymptotic Performance*: The final learned performance of an agent in the target task may be improved via transfer.

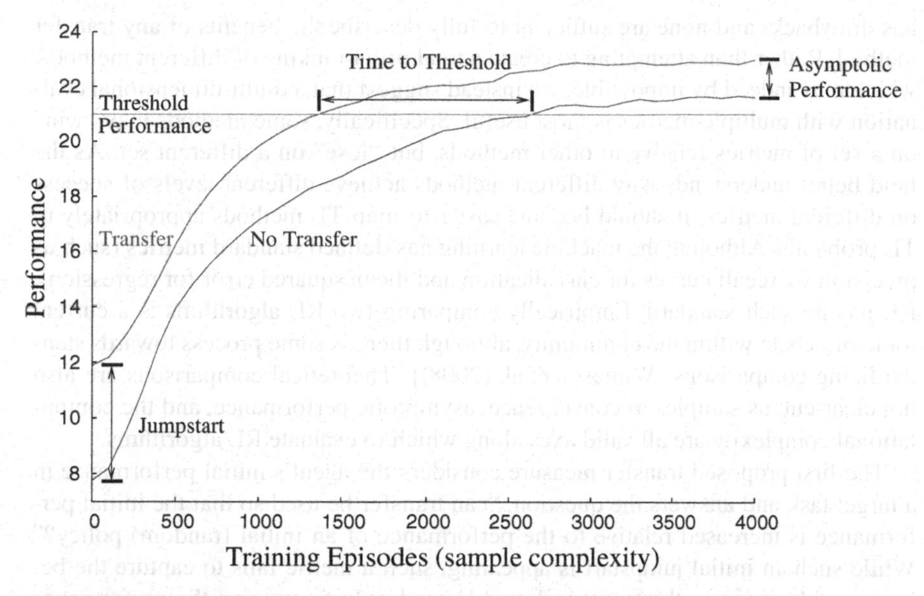

Fig. 1.2 Many different metrics for measuring TL are possible. This graph show benefits to the jumpstart, asymptotic performance, time to threshold, and total reward (the area under the learning curve).

3. *Total Reward*: The total reward accumulated by an agent (i.e., the area under the learning curve) may be improved if it uses transfer, compared to learning without transfer.
4. *Transfer Ratio*: The ratio of the total reward accumulated by the transfer learner and the total reward accumulated by the non-transfer learner.
5. *Time to Threshold*: The learning time needed by the agent to achieve a pre-specified performance level may be reduced via knowledge transfer.

Metrics 1–4 are most appropriate in the fully autonomous scenario as they do not charge the agent for time spent learning any source tasks. To measure the total time, the metric must account for time spent learning one or more source tasks, which is natural when using metric 5. Other metrics have been proposed in the literature, but we choose to focus on these five because they are sufficient to describe the methods surveyed in this monograph.

We may think of learning time as a surrogate for *sample complexity*. Sample complexity (or data complexity) in RL refers to the amount of data required by an algorithm to learn. It is strongly correlated with learning time because RL agents only gain data by collecting it through repeated interactions with an environment.

1.2.1 Empirical Transfer Comparisons

The previous section enumerated five possible TL metrics, and while others are possible, these represent the methods most commonly used. However, each metric has drawbacks and none are sufficient to fully describe the benefits of any transfer method. Rather than attempting to create a total order ranking of different methods, which may indeed by impossible, we instead suggest that a multi-dimensional evaluation with multiple metrics is most useful. Specifically, some methods may "win" on a set of metrics relative to other methods, but "lose" on a different set. As the field better understands why different methods achieve different levels of success on different metrics, it should become easier to map TL methods appropriately to TL problems. Although the machine learning has defined standard metrics (such as precision vs. recall curves for classification and mean squared error for regression), RL has no such standard. Empirically comparing two RL algorithms is a current topic of debate within the community, although there is some process towards standardizing comparisons [Whiteson et al. (2008)]. Theoretical comparisons are also not clear-cut, as samples to convergence, asymptotic performance, and the computational complexity are all valid axes along which to evaluate RL algorithms.

The first proposed transfer measure considers the agent's initial performance in a target task and answers the question, "can transfer be used so that the initial performance is increased relative to the performance of an initial (random) policy?" While such an initial jumpstart is appealing, such a metric fails to capture the behavior of *learning* in the target task and instead only focuses on the performance before learning occurs.

Asymptotic performance, the second proposed metric, compares the final performance of learners in the target task both with and without transfer. However, it may be difficult to tell when the learner has indeed converged (particularly in tasks with infinite state spaces) or convergence may take prohibitively long. In many settings the number of samples required to learn is most critical, not the performance of a learner with an infinite number of samples. Further, it is possible for different learning algorithms to converge to the same asymptotic performance but require very different numbers of samples to reach the same performance.

A third possible measure is that of the total reward accumulated during training. Improving initial performance and achieving a faster learning rate will help agents accumulate more on-line reward. RL methods are often not guaranteed to converge with function approximation and even when they do, learners may converge to different, sub-optimal performance levels. If enough samples are provided to agents (or, equivalently, learners are provided sufficient training time), a learning method which achieves a high performance relatively quickly will have less total reward than a learning method which learns very slowly but eventually plateaus at a slightly higher performance level. This metric is most appropriate for tasks that have a well-defined duration.

A fourth measure of transfer efficacy is that of the ratio of areas defined by two learning curves. Consider two learning curves in the target task where one uses transfer and one does not. Assuming that the transfer learner accrues more reward, the area under the transfer learning curve will be greater than the area under the non-transfer learning curve. The ratio

$$r = \frac{\textit{area under curve with transfer - area under curve without transfer}}{\textit{area under curve without transfer}}$$

gives a metric that quantifies improvement from TL. This metric is most appropriate if the same final performance is achieved, or there is a predetermined time for the task. Otherwise the ratio will directly depend on how long the agents act in the target task.

While such a metric may be appealing as a candidate for inter-task comparisons, we note that the transfer ratio is not scale invariant. For instance, if the area under the transfer curve were 1000 units and the area under the non-transfer curve were 500, the transfer ratio would be 1.0. If all rewards were multiplied by a constant, this ratio would not change. But if an offset were added (e.g., each agent is given an extra +1 at the end of each episode, regardless of the final state), the ratio would change. The evaluation of a TL algorithm with the transfer ratio is therefore closely related to the reward structure of the target task being tested. Lastly, we note that although none of the papers discussed in this monograph use such a metric, we hope that it will be used more often in the future.

The final metric, Time to Threshold, suffers from having to specify a, potentially arbitrary, performance agents must achieve. While there have been some suggestions how to pick such thresholds appropriately [Taylor et al. (2007a)], the relative benefit of TL methods will clearly depend on the exact threshold chosen, which will necessarily be domain- and learning method-dependant. While choosing a range of thresholds to compare over may produce more representative measures (c.f., [Taylor et al. (2007b)]), this leads to having to generating a time vs. threshold curve rather than producing a single real valued number that evaluates a transfer algorithm's efficacy.

A further level of analysis that could be combined with any of the above methods would be to calculate a ratio comparing the performance of a TL algorithm with that of a human learner. For instance, a set of human subjects could learn a given target task with and without having first trained on a source task. By averaging over their performances, different human transfer metrics could be calculated and compared to that of a TL algorithm. However, there are many ways to manipulate such a meta-metric. For instance, if a target task is chosen that humans are relatively proficient at, transfer will provide them very little benefit. If that same target task is difficult for a machine learning algorithm, it will be relatively easy to show that the TL algorithm is quite effective relative to human transfer, even if the agent's absolute performance is extremely poor.

A major drawback of all the metrics discussed is that none are appropriate for inter-domain comparisons. Developing fair metrics that apply across multiple problem domains would be very useful and allow better comparisons of methods. Such inter-domain metrics may be infeasible in practice, in which case standardizing on a set of test domains would assist in comparing different TL methods (as discussed further in Section 9.4.2). In the absence of either a set of inter-domain metrics or a standard benchmark suite of domains, we limit our comparisons of different TL methods in this monograph to their applicability, assumptions, and algorithmic differences. When discussing different methods, we may opine on the method's relative performance, but we remind the reader that such commentary is largely based on intuition rather than empirical data.

1.2.2 Dimensions of Comparison

In addition to differing on evaluation metrics, we categorize TL algorithms along five dimensions, which we use as the main organizing framework for our survey of the literature:

I *Task difference assumptions*: What assumptions does the TL method make about how the source and target are allowed to differ? Examples of things that can differ between the source and target tasks include different system dynamics (i.e., the target task becomes harder to solve is some incremental way), or different sets of possible actions at some states. Such assumptions define the types of source and target tasks that the method can transfer between. Allowing transfer to occur between less similar source and target tasks gives more flexibility to a human designer in the human-guided scenario. In the fully autonomous scenario, more flexible methods are more likely to be able to successfully apply past knowledge to novel target tasks.

II *Source task selection*: In the simplest case, the agent assumes that a human has performed source task selection (the human-guided scenario), and transfers from one or more selected tasks. More complex methods allow the agent to select a source task or set of source tasks. Such a selection mechanism may additionally be designed to guard against *negative transfer*, where transfer hurts the learner's performance. The more robust the selection mechanism, the more likely it is that transfer will be able to provide a benefit. While no definitive answer to this problem exists, successful techniques will likely have to account for specific target task characteristics. For instance, [Carroll and Seppi (2005)] motivate the need for general task similarity metrics to enable robust transfer, propose three different metrics, and then proceed to demonstrate that none is always "best," just as there is never a "best" inductive bias in a learning algorithm.

III *Task Mappings*: Many methods require a mapping to transfer effectively: in addition to knowing that a source task and target task are related, they need to know *how* they are related. *Inter-task mappings* (discussed in detail later in Section 3.1.3) are a way to define how two tasks are related. If a human is in the loop, the method may assume that such task mappings are provided; if the agent is expected to transfer autonomously, such mappings have to be learned. Different methods use a variety of techniques to enable transfer, both on-line (while learning the target task) and offline (after learning the source task but before learning the target task). Such learning methods attempt to minimize the number of samples needed and/or the computational complexity of the learning method, while still learning a mapping to enable effective transfer.

IV *Transferred Knowledge*: What type of information is transferred between the source and target tasks? This information can range from very low-level information about a specific task (i.e., the expected outcome when performing an action in a particular location) to general heuristics that attempt to guide learning. Different types of knowledge may transfer better or worse depending on task similarity. For instance, low-level information may transfer across closely related tasks, while high-level concepts may transfer across pairs of less similar tasks. The mechanism that transfers knowledge from one task to another is closely related to what is being transferred, how the task mappings are defined (III), and what assumptions about the two tasks are made (I).

V *Allowed Learners*: Does the TL method place restrictions on what RL algorithm is used, such as applying only to temporal difference methods? Different learning algorithms have different biases. Ideally an experimenter or agent would select the RL algorithm to use based on characteristics of the task, not on the TL algorithm. Some TL methods require that the source and target tasks be learned with the same method, other allow a class of methods to be used in both tasks, but the most flexible methods decouple the agents' learning algorithms in the two tasks.

1.3 Transfer for Reinforcement Learning

In this section we first give a brief overview of notation. We then summarize the methods discussed in this monograph using the five dimensions previously discussed, as well as enumerating the possible attributes for these dimensions. Lastly, learning paradigms with goals similar to transfer are discussed in Section 3.1.4.

1.3.1 Reinforcement Learning Notation

RL problems are typically framed in terms of *Markov decision processes* (MDPs) [Puterman (1994)]. For the purposes of this monograph, *MDP* and *task* are used interchangeably. In an MDP, there is some set of possible perceptions of the current *state* of the world, $s \in S$, and a learning agent has one or more initial starting states, $s_{initial}$. The *reward function*, $R : S \mapsto \mathbb{R}$, maps each state of the environment to a single number which is the instantaneous reward achieved for reaching the state. If the task is *episodic*, the agent begins at a start state and executes actions in the environment until it reaches a terminal state (one or more of the states in s_{final}, which may be referred to as a *goal state*), at which point the agent is returned to a start state. An agent in an episodic task typically attempts to maximize the average reward per episode. In non-episodic tasks, the agent attempts to maximize the total reward, which may be discounted. By utilizing a discount factor, γ, the agent can weigh immediate rewards more heavily than future rewards, allowing it to maximize a non-infinite sum of rewards.

An agent knows its current state in the environment, $s \in S$. The agent's observed state may be different from the true state if there is perceptual noise. If the agent only receives *observations* and does not know the true state, the agent may treat approximate its true state as the observation (c.f., [Stone et al. (2005)]), or it may learn using the Partially Observable Markov Decision Process (POMDP) (c.f., [Kaelbling et al. (1998)]) problem formulation, which is beyond the scope of this monograph. TL methods are particularly relevant in MDPs that have a large or continuous state, as these are the problems which are slow to learn *tabula rasa* and for which transfer may provide substantial benefits. Such tasks typically factor the state using *state variables* (or *features*), so that $s = \langle x_1, x_2, \ldots, x_n \rangle$. The set A describes the *actions* available to the agent, although not every action may be possible in every state. The *transition function*, $T : S \times A \mapsto S$, takes a state and an action and returns the state of the environment after the action is performed. Transitions may be non-deterministic, making the transition function a probability distribution function. A learner senses the current state, s, typically knows A, and what state variables comprise S; however, it is generally not given R or T. An agent may choose to directly learn a *policy*, $\pi : S \mapsto A$, which fully defines how a learner interacts with the environment. Another common technique is to estimate an *action-value function*, $Q : S \times A \mapsto \mathbb{R}$, where $Q(s,a)$ is the expected return found when executing action a from state s, and greedily following the current policy thereafter.

1.3.2 What Follows

This chapter has introduced transfer learning and the goals of this monograph. The next chapter familiarizes the reader with the RL framework, the base learning algorithms used, and metrics to quantify the effects of transfer. Although this monograph does not focus on improving base RL algorithms, they are important to the clarity of later TL experiments. Related work is presented in Chapter 3 with an emphasis on how methods in this monograph differ from existing approaches. The four different domains used in this monograph are presented in Chapter 4, as well as learning results for the base RL algorithms. Each of these domains is chosen to highlight or test a different aspect of the transfer learning algorithms.

Chapter 5 defines inter-task mappings and introduces Value Function Transfer, the first TL method, which works by using Q-values learned in a source task to initialize Q-values in target task agents. While effective, Value Function Transfer requires that both the source and target task agents use TD algorithms, and requires that both agents have the same type of function approximation (i.e., learning representation).

Chapter 6 introduces two methods that reduce the requirements imposed by Value Function Transfer. The first, Q-Value Reuse, again requires that both the source task and target task agents use TD learning methods, but allows source and target task agents to use different types of function approximation. The second, Policy Transfer, enables transfer between direct policy search methods, a class of RL methods distinct from TD methods.

Chapter 7 discusses three TL methods that allow the source task and target task agents to use different learning methods. Again, all methods are compatible with inter-task mappings. The first, TIMBREL, allows any type of learning algorithm in the source task and then improves a model-learning RL target task agent. Rule Transfer also allows any type of learning in the source task and can be applied to a

TD learning agent in the target task. Third, Representation Transfer is a set of algorithms that provide even more flexibility than the previous methods. Representation Transfer allows the source agent and target agent to differ by learning method, function approximator, or the function approximator's parameterization, in addition to transferring between different tasks.

Chapter 8 presents two methods for learning inter-task mappings. The first, Mapping Learning via Classification, uses a classification algorithm to discover similarities between pairs of tasks based on observed action effects. The second, MASTER, leverages regression to learn an approximate transition model and test different possible mappings by measuring the error between outcomes predicted by the model and outcomes observed in data. MASTER is particularly significant because it relies on no domain knowledge and thus can be incorporated into a fully autonomous TL agent.

Chapter 9 concludes with a summary of the monograph and possible future directions.

Chapter 2
Reinforcement Learning Background

This monograph focuses on transfer learning in reinforcement learning domains; some RL background is necessary. Our goal in this chapter is to briefly discuss RL concepts and notation used in this monograph so that the reader may understand later TL algorithms and experiments. Readers who desire a more comprehensive treatment of the reinforcement learning framework are referred to [Kaelbling et al. (1996)] and [Sutton and Barto (1998)].

2.1 Framing the Reinforcement Learning Problem

RL problems are typically framed in terms of *Markov decision processes* [Puterman (1994)] (MDPs). For the purposes of this monograph, "MDP" and "task" can be used interchangeably. An MDP is specified by the 4-tuple $\langle S, A, T, R \rangle$.[1] S is the set of *states* in the task. The available actions are enumerated in the set A, although not every action may be possible in every state. The *transition function*, $T : S \times A \mapsto S$, takes a state and an action and returns the state of the environment after the action is performed. Transitions may be non-deterministic, making the transition function a probability distribution function. The *reward function*, $R : S \mapsto \mathbb{R}$, maps each state of the environment to a real-valued number which is the instantaneous reward achieved for reaching the state.

A learning agent senses its current state $s \in S$. The agent's observed state may be different from the true state if there is perceptual noise. If the task is *episodic*, the agent begins at a start state, $s_{initial}$, and executes actions in the environment until it reaches a terminal state, s_{final}, at which point it is returned to a start state. In some tasks where the agent is given no reward except when reaching a terminal state, it is convenient to think of the final states as *goal* states. An agent in an episodic task typically attempts to maximize the average reward per episode. In non-episodic tasks, the agent attempts to maximize the total reward, which may be discounted.[2]

[1] Some formulations also explicitly include a start state distribution, S_0, and a terminal state distribution, S_f.

[2] By utilizing a discount factor, γ, the agent can weigh immediate rewards more heavily than future rewards, allowing it to maximize the expectation of an infinite sum of rewards.

M.E. Taylor: Transfer in Reinforcement Learning Domains, SCI 216, pp. 15–29.
springerlink.com

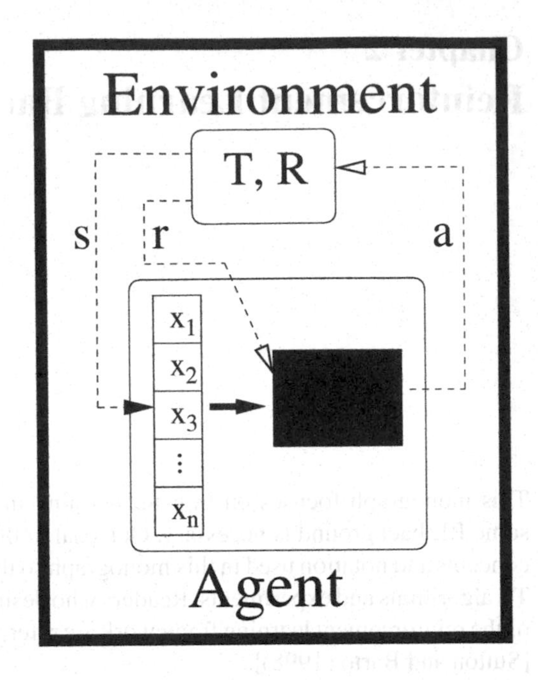

Fig. 2.1 An agent interacts with an environment by sequentially selecting an action in an observed state, with the objective of maximizing an external reward signal.

Transfer learning methods are particularly relevant in MDPs that have a large or continuous state space, as these are the problems which are slow to learn *tabula rasa* and for which transfer may provide substantial benefits. Such tasks typically factor the state using *state variables*, so that $s = \langle x_1, x_2, \ldots, x_n \rangle$ (see Figure 2.1).

A *policy*, $\pi : S \mapsto A$, fully defines how a learner interacts with the environment by mapping perceived environmental states to actions. The success of an agent is determined by how well it maximizes the total reward it receives in the long run while acting under some policy π. An *optimal policy*, $\pi^\star$, is a policy which does maximize the expectation of this value. Any reasonable learning algorithm attempts to modify π over time so that the agent's performance approaches that of $\pi^\star$ in the limit.

Rather than computing a policy directly, many learning methods first estimate an *action-value function*, $Q : S \times A \mapsto \mathbb{R}$. $Q(s,a)$ is the expected *return* (or total reward) found when executing action a from state s, and then greedily following the current policy thereafter. The current policy may be generated from Q by simply selecting the action that has the highest value for the current state. Another possibility is to calculate the *value function*, $V : S \mapsto \mathbb{R}$, which maps states to the expected return. Value functions are typically learned when the agent has a model of the task: to generate a policy, the agent calculates $V(s')$ for all possible s', which can only be done by knowing $T(s,a)$ for $a \in A$.

There are many possible approaches for learning a policy or action-value function, which will be discussed in Section 2.3. We first provide an overview of *function approximation* for RL, which will be crucial for learning in large tasks.

2.2 Function Approximation

In tasks with small and discrete state spaces, π, Q, and T can be represented in a table. As the state space grows, using a table becomes impractical, or impossible if the state space is continuous. In this monograph we use four function approximators and show that all are capable of successfully utilizing transfer. Specifically, we consider linear tile-coding function approximation, also known as *cerebellar model arithmetic computers* (CMACs), which has been successfully used in many reinforcement learning systems [Albus (1981)]; *radial basis functions* (RBFs) [Sutton and Barto (1998)], a continuous variant of CMACs; *artificial neural networks* (ANNs), a biologically-inspired method for computing with a network simple computing units; and an instance-based approximator, which stores observed data to predict future outcomes.

All of the function approximators have parameters that must be set to accurately reflect the underlying task. Although some work in RL [Dean and Givan (1997), Li et al. (2006), Mahadevan and Maggioni (2007)] has taken a more systematic approaches to *state abstractions* (also called *structural abstractions*), the majority of current research relies on humans to help bias a learning agent by carefully selecting a function approximator with parameters appropriate for a given task.

2.2.1 Cerebellar Model Arithmetic Computers

CMACs take arbitrary groups of continuous state variables and lay infinite, axis-parallel tilings over them (see Figure 2.2). This allows discretization of continuous state space into tiles while maintaining the capability to generalize via multiple overlapping tilings. Increasing the tile widths allows better generalization; increasing the number of tilings allows more accurate representations of smaller details. The number of tiles and the width of the tilings are generally handcoded: this sets the center, c_i, of each tile and dictates which state values will activate which tiles. The function approximation is trained by changing how much each tile contributes to the output of the function approximator (see Figure 2.3). Thus, the output from the CMAC is the computed sum:

$$f(x) = \sum_i w_i f_i(x) \tag{2.1}$$

but only tiles which are activated by the current state features contribute to the sum:

$$f_i(x) = \begin{cases} 1, & \text{if tile } i \text{ is activated} \\ 0, & \text{otherwise} \end{cases}$$

Unless otherwise specified, in this monograph all weights in a CMAC are initialized to zero. However, if information about the task is known in advance, a more informed initial weight selection could be used. Note that although the majority of experiments in this monograph will use one-dimensional tilings (one per state variable), the principles above apply in the n-dimensional case.

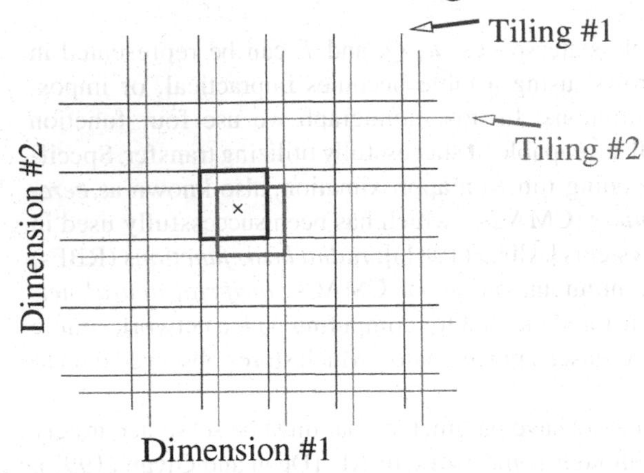

Fig. 2.2 A CMAC's value is computed by summing the weights, w_i, from multiple activated tiles (outlined above with thicker lines). State variables are used to determine which tile is activated in each of the different tilings.

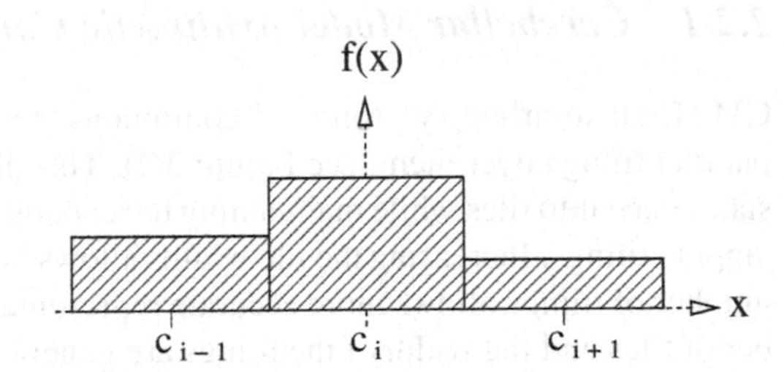

Fig. 2.3 A CMAC approximator computes its output via a weighted sum of step functions, where the contribution of the ith step function is controlled by the weight w_i. This figure depicts the output of a 1-dimensional CMAC with a single tiling.

2.2.2 Radial Basis Functions

We utilize RBFs to generalize the tile coding idea to continuous functions. When considering a single state variable, an RBF approximator is a linear function approximator

$$f(x) = \sum_i w_i f_i(x) \tag{2.2}$$

where the basis functions have the form

$$f_i(x) = \phi(|x - c_i|) \tag{2.3}$$

x is the value of the current state variable, c_i is the center of feature i (similar to the CMAC, Equation 2.1), and w_i represents weights that can be modified over time by a learning algorithm. Here we set the features to be evenly spaced Gaussian radial basis functions, where

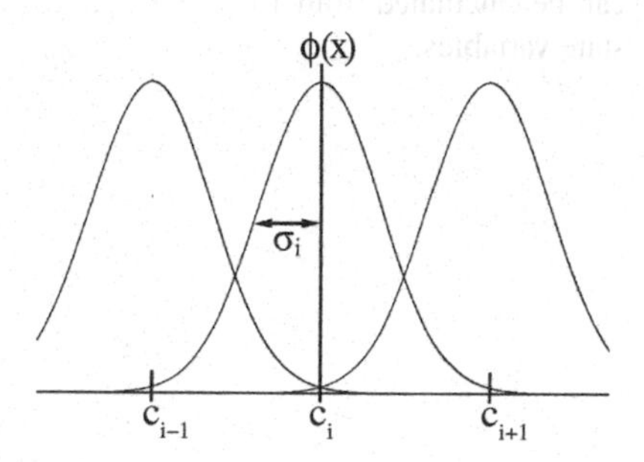

Fig. 2.4 An RBF approximator computes Q(s,a) via a weighted sum of Gaussian functions, analogous to the step weights in Figure 2.3. The contribution from the ith Gaussian is weighted by the distance from its center, c_i, to the relevant state variable, and by the learned parameter w_i. σ can be tuned to control the width of Gaussians and thus how much the function approximator generalizes.

$$\phi(x) = \exp(-\frac{x^2}{2\sigma^2}) \tag{2.4}$$

(see Figure 2.4). The σ parameter controls the width of the Gaussian function and therefore the amount of generalization over the state space.

As with CMACs, experiments in this monograph using RBFs are primarily 1-dimensional, with a different RBF tiling used for each state variable. When RBFs are used to approximate the action-value function, Equations 2.2-2.4 are used to calculate Q-values for (s,a) pairs. All weights are initially set to zero, but over time learning methods change the values of the weights so that the resulting Q-values more closely predict the true returns and thus improve the policy implicitly defined by Q.

2.2.3 Artificial Neural Networks

The ANN function approximator similarly allows a learner to approximate the action-value function, given a set of continuous, real valued, state variables. Although ANNs have been shown difficult to train in certain situations on relatively simple RL problems [Boyan and Moore (1995), Pyeatt and Howe (2001)], they have had notable successes on some RL tasks [Tesauro (1994), Crites and Barto (1996), Whiteson and Stone (2006)]. Each input to the ANN is set to the value of a state variable and each output corresponds to an action. Activations of the output nodes correspond to Q-values (see Figure 2.5 for a diagram).

When using ANNs to approximate an action-value function, we use non-recurrent feedforward networks. Each node in the input layer is given the value of a different state variable and each output node corresponds is the calculated Q-value for a different action.[3] The number of inputs and outputs are thus determined by the task's specification, but the number of hidden nodes is specified by the agent's designer.

[3] Alternatively, there could be $|A|$ different neural networks, each with one output node, that corresponding to $Q(\cdot,a)$.

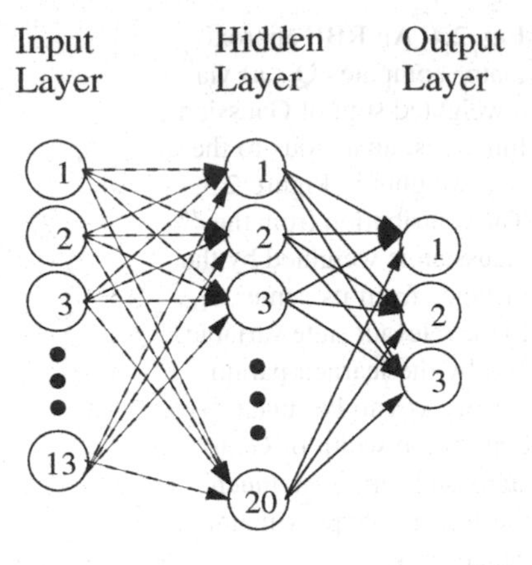

Fig. 2.5 This diagram of an artificial feedforward 13-20-3 network shows how Q-values for three actions can be calculated from 13 state variables.

Algorithm 1. ANNOutput(*Topology*, *Weights*, *Inputs*, *out*)

1: # *Topology* is a data structure that contains the number of nodes in each layer
2: # *Weights* is a data structure that contains the values of all weights in the network
3: # *Inputs* is a vector describing the current state, scaled from $[-1,1]$
4: # *out* is the index of the output node
5: **for** $j \in$ `Topology.getHiddenNodes()` **do**
6: *HiddenNodeSum[j]* $\leftarrow 0$
7: **for** $i \in$ `Topology.getInputs()` **do**
8: *HiddenNodeSum[j]* $\leftarrow$ *HiddenNodeSum[j]* + *Inputs[i]* $\times$ *Weights.getLayer1(i,j)*
9: *HiddenNodeSum[j]* $\leftarrow \frac{1.0}{1.0+e^{-HiddenNodeSum[j]}}$ # sigmoid function
10: *Output* $\leftarrow 0$
11: **for** $j \in$ `Topology.getHiddenNodes()` **do**
12: *Output* $\leftarrow$ *Output* + *HiddenNodeSum[j]* $\times$ *Weights.getLayer2(j,out)*
13: **return** *Output* # Output node is linear (no sigmoid function)

Note that by accepting multiple inputs the neural network can determine its output by considering multiple state variables in conjunction (as opposed to a CMAC or RBF consisting of only 1-dimensional tilings). Nodes in the hidden layer have a sigmoid transfer function and output nodes are linear. Weights for connections in the network are typically initialized to random values near zero. Algorithm 1 details the process of calculating outputs for such a fully-connected two-layer neural network. The networks are trained using standard backpropagation, where the error signal to modify weights is generated by the learning algorithm (see Section 2.3.1), as with the other function approximators.

ANNs can also be used to specify a policy directly, where the network functions as an *action selector*. In this case, the number of inputs and outputs will again be fixed per the MDP. Instead of calculating Q-values at the output nodes, the agent can follow the greedy policy by selecting the action corresponding to the output with the highest activation. This monograph utilizes ANNs for action selection when learning with direct policy search (see Section 2.3.2). When calculating the output of an ANN that uses NEAT, the procedure is more complex due to the possibility of complex network topologies, such as recurrent links between layers. However, the calculation is conceptually similar: node values are multiplied by weights as they propagate through the network and the sums are then modified by sigmoid functions.

2.2.4 Instance-Based Approximation

CMACs, RBFs, and ANNs aim to represent a complex function with a relatively small set of parameters that can be changed over time. In this section we discuss instance-based approximation, which stores instances experienced by the agent (i.e., $\langle s,a,r,s'\rangle$) to predict the underlying structure of the environment. Specifically, this approximation method will be used by a model-learning method (see Section 2.3.3), which learns to approximate T and R by observing the agent's experience when interacting with an environment.

Consider the case where an agent is acting is a discrete environment with a small state space. The agent could record every instance that it experienced in a table. If the transition function were deterministic, as soon as the agent observed every possible (s,a) pair, it could calculate the optimal policy. If the transition function were instead stochastic, the agent would need to take multiple samples for every (s,a) pair. Given a sufficient number of samples, as determined by the variance in the resulting r and s', the agent could again directly calculate the optimal policy (Section 2.3 will briefly discuss dynamic programming, one such way of calculating $\pi^\star$ in this situation).

When used to approximate T and R for tasks with continuous state spaces, using instances for function approximation becomes significantly more difficult (see Figure 2.6). In a stochastic task the agent is unlikely to ever visit the same state twice, with the possible exception of a start state, and thus approximation is critical. Furthermore, since one can never gather "enough" samples for every (s,a) pair,

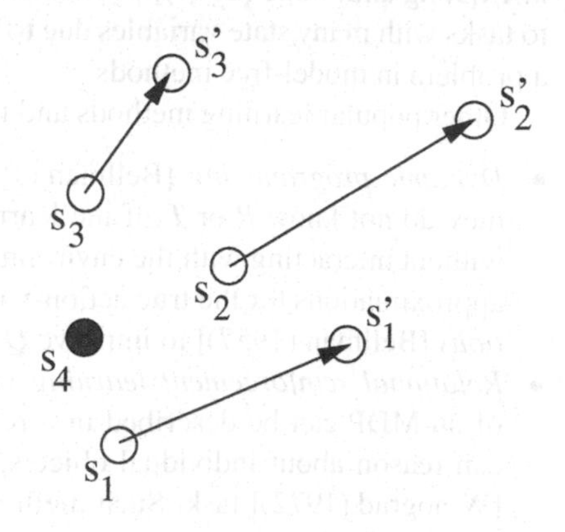

Fig. 2.6 Consider an MDP with only a single action a. In one part of the state space, three transitions have been recorded. The instance-based approximator must now estimate s'_4 from these instances, which is equivalent to estimating $T(s_4,a)$.

such methods generally need to determine which instances are necessary to store so that the memory requirements are not unbounded. In Section 2.3.3 we will discuss one such model-learning method that uses instance-based approximation to learn in continuous state spaces.

2.3 Learning Methods

This section discusses different approaches to learning policies in MDPs. We first give a high-level overview of some of the more popular methods and then detail the four learning methods used in this monograph's experiments.

Temporal difference (TD) [Sutton and Barto (1998)] methods, such as *Q-learning* [Sutton (1988), Watkins (1989)] and *Sarsa* [Rummery and Niranjan (1994), Singh and Sutton (1996)], learn by backing up experienced rewards through time. Their goal is to learn to approximate an action-value function, from which a policy is easily derived. TD methods are among the most popular due to their relative simplicity, theoretical guarantees, and empirical successes. Sarsa will be discussed in detail in Section 2.3.1.

Policy search [Williams (1992), Williams and Singh (1999), Ng and Jordan (2000), Baxter and Bartlett (2001)] methods are in some sense simpler then TD methods. They directly modify a policy over time to increase the expected long-term reward by using search or other optimization techniques. *Neuro Evolution of Augmenting Topologies* [Stanley and Miikkulainen (2002)], the policy search method used in this monograph, is discussed in Section 2.3.2.

Model-based methods [Moore and Atkeson (1993), Kearns and Singh (1998)] (also known as or *Model-learning* methods) attempt to estimate the true model of the environment (i.e., T and R) by interacting with the environment over time. *Instance based methods* [Ormoneit and Sen (2002)] save observed interactions with the environment and leverage the instance to directly predict the model. The learned model is then typically used to help the agent decide how to efficiently explore or plan trajectories so that it can accrue higher rewards. While very successful in small tasks, few such methods handle continuous state spaces (see Fitted R-MAX[Jong and Stone (2007)] in Section 2.3.3), and they may have trouble scaling to tasks with many state variables due to the "curse of dimensionality," which is also a problem in model-free methods.

Other popular learning methods and techniques include:

- *Dynamic programming* [Bellman (1957)]: While most algorithms assume that they do not know R or T, if the learner is supplied these functions, it can learn without interacting with the environment. Instead, the agent iteratively computes approximations for the true action-value function, by solving the *Bellman Equations* [Bellman (1957)] to improve Q or V (and the resulting policy) over time.
- *Relational reinforcement learning* (RRL) [Dzeroski et al. (2001)]: If the state of an MDP can be described in a relational or first-order language, algorithms can reason about individual objects, such as a single block in a *Blocks World* [Winograd (1972)] task. Such methods may simplify transfer, as the number of

objects can change in a task and the learned action-value function can be applied without modification.

- *Bayesian RL* [Dearden et al. (1998)]: After specifying a prior, which may be uniform, a Bayesian mathematical model can be used to explicitly represent uncertainty in the components of the model, updating expectations over time.
- *Batch methods*: On-line methods require that the agent update its knowledge as it interacts with the environment. Batch, or offline, methods are designed to be more sample efficient, as they can store environmental interaction data and use the set multiple times to learn to approximate Q or π. Additionally, such methods allow a clear separation of the learning mechanism from the exploration mechanism (which much decide whether to attempt to gather more data about the environment or exploit the current best policy).
- *Temporal abstractions*: Rather than learn over the actions in the MDP, different methods have been designed to group series of actions together. These macro-actions or *options* [Sutton et al. (1999)] may allow the agent to leverage the sequence of actions to learn its task with less data. Options may be hand specified or learned.
- *Hierarchical methods*: If a task can be decomposed into different sub-tasks, methods such as in MAXQ [Dietterich (2000)] can exploit such a hierarchy. By subdividing the MDP in this way, an agent can learn each subtask relatively quickly and then combine them together, resulting in an overall learning speed improvement relative to "flat" learning methods.

2.3.1 Sarsa

The experiments presented in this monograph use Sarsa as a representative TD method. Sarsa is an appropriate choice because of past empirical successes [Sutton (1996), Sutton and Barto (1998), Stone et al. (2005)]. Sarsa learns to estimate $Q(s,a)$ by backing up the received rewards through time. Sarsa is an acronym for State Action Reward State Action, describing the 5-tuple needed to perform the update, $(s_t, a_t, r, s'_t, a'_t)$, where s_t is the agent's current state at time t, a_t is the agent's action, r is the immediate reward the agent receives from the environment, s'_t is the agent's next state, and a'_t is the agent's subsequently chosen action. After each action, Q-values are updated according to the following rule:

$$Q(s_t, a_t) \leftarrow (1-\alpha)Q(s_t, a_t) + \alpha(r + Q(s'_t, a'_t)) \tag{2.5}$$

where α is the learning rate. The full algorithm is detailed in Algorithm 2. Note that if the task is non-episodic (continuing) we need to apply an extra discount factor γ in the range (0,1) to $Q(s'_t, a'_t)$ so that the future rewards are weighted less than the immediate rewards.

Like other TD methods, Sarsa estimates the value of a given state-action pair by bootstrapping off the estimates of other such pairs. In particular, the value of a given state-action pair (s_t, a_t) can be estimated as $r + Q(s'_t, a'_t)$, which is the immediate reward received during the transition plus the value of the subsequent state-action pair.

Algorithm 2. Sarsa(α, A)

```
1: # α is the learning rate
2: # A is the set of possible actions
Require: Q has been initialized (arbitrarily)
3: loop
4:   Observe state s
5:   Select action a by evaluating Q(s,·)
6:   repeat
7:     Execute a
8:     Observe r and s'
9:     Select a' by evaluating Q(s',·)
10:    Q(s,a) ← (1 − α)Q(s,a) + α(r + Q(s',a'))
11:    s ← s', a ← a'
12:  until s is a terminal state        # the episode ends
```

Sarsa's update rule thus takes the old action-value estimate, $Q(s_t, a_t)$, and moves it incrementally closer towards this new estimate. The learning rate parameter, α, controls the size of these increments. These action-value estimates should become more accurate over time and therefore improve the agent's performance. When acting in small, discrete environments, the agent can store Q-values in a table. In large MDPs or MDPs with continuous state variables, a function approximator (e.g., a CMAC, RBF, or ANN) is used to calculate a Q-value for a given (s,a) pair.

The intuition behind this update rule is that over time the Q-values will converge towards the true values through many small "backups." In episodic tasks, the Q-value of the goal state is defined by the task and thus the final Q-value of an episode will have the correct value once reached. Over time the correct values are "backed up" so that, ideally, the correct values for all regularly visited states will be learned.

Exploration, when the agent chooses an action to learn more about the environment, must be balanced with *exploitation*, when the agent selects what it believes to be the best action. One simple approach to action selection (lines 5 and 9 in Algorithm 2) is ε*-greedy* action selection: the agent selects a random action with chance ε, and the current best action is selected with probability $1-\varepsilon$ (where ε is in [0,1]). See Algorithm 3 for an algorithmic description. Over time, the agent may decay ε to encourage more exploitation as learning progresses.

One useful refinement to Sarsa utilizes *eligibility trances* [Watkins (1989)]. In brief, traces are used to keep track of (s,a) pairs that have been visited recently. Rather than doing a single update at each timestep, recently visited (s,a) pairs also share some of the update because they are partially "responsible" for the agent's current situation. New (s,a) pairs are set to have an eligibility of 1 and on each update all eligibilities are decayed by a fixed parameter, typically denoted λ. The Sarsa(λ) update [Rummery and Niranjan (1994)] is:

$$Q(s_t,a_t) \leftarrow (1-\alpha)Q(s_t,a_t) + \alpha e(s_t,a_t)(r + Q(s'_t,a'_t))$$

Algorithm 3. SelectAction(Q, A, s, ε)

```
1: # Q is the action-value function
2: # A is the set of actions available to the agent
3: # s is the current state
4: # ε is the exploration rate
5: if GetRand(0,1) ≤ ε then
6:    return SelectRandom(A)
7: else
8:    return argmax_a(Q(s,a))        # (for a ∈ A)
```

where e is the eligibility trace:

$$e_t(s,a) \begin{cases} 1 & \text{if } s = s_t \text{ and } a{=}a_t \\ \lambda e_{t-1}(s,a) & \text{otherwise} \end{cases}$$

2.3.2 *NeuroEvolution of Augmenting Topologies (NEAT)*

NeuroEvolution of Augmenting Topologies [Stanley and Miikkulainen (2002)] (NEAT) is used in this monograph as a representative policy search method for RL. NEAT belongs to the class of *genetic algorithms* which use biologically-inspired evolutionary techniques to search the policy space, and is an appropriate choice because of it has had a number of empirical successes on RL tasks, such as pole balancing [Stanley and Miikkulainen (2002)], game playing [Stanley and Miikkulainen (2004b)], robot control [Stanley and Miikkulainen (2004a)], and Server Job Scheduling [Whiteson and Stone (2006)]. NEAT utilizes ANNs, but rather than computing Q-values, it directly represents a policy via the network's topology and weights, both of which are modified over time to improve performance.

Most neuroevolutionary [Yao (1999)] systems require the network topology to be fixed and given (i.e., how many hidden nodes there are and how they are connected). By contrast, NEAT automatically evolves the topology by combining the search for network weights with evolution of the network structure. In NEAT, a population of genomes, each of which describes a single neural network, is evolved over time: each genome is evaluated and the fittest individuals reproduce through crossover and mutation.

NEAT begins with a population of simple networks with no hidden nodes and inputs connected directly to sigmoidal outputs. Two special mutation operators introduce new structure incrementally, as depicted in Figure 2.7. Only structural mutations that improve performance tend to survive evolution and thus NEAT often finds an appropriate level of complexity needed for a given problem.

Since NEAT is a general purpose optimization technique, it can be applied to a wide variety of problems. In this monograph, we use NEAT for policy search RL. Each neural network in the population represents a candidate policy in the form of an action selector. The inputs to the network describe the agent's current state.

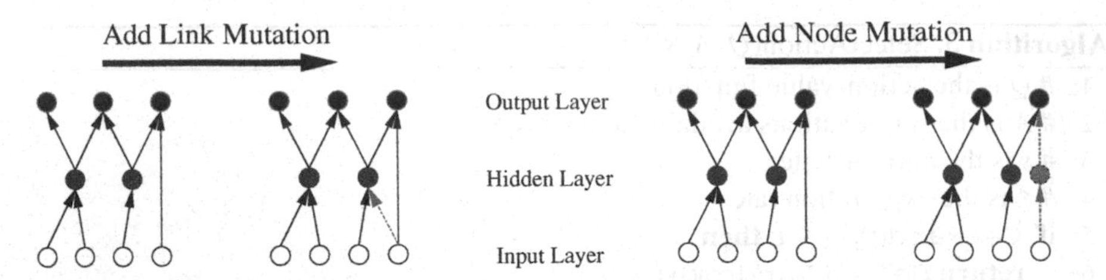

Fig. 2.7 Examples of NEAT's mutation operators for adding structure to networks. Left: a link, shown as a dotted red line, is added to connect two nodes. Right: a hidden node is added by splitting a link in two.

Algorithm 4. NEAT(*numInputs, numOutputs, popSize, e, NEATParams*)

1: # *numOutputs* is the required number of input nodes
2: # *numInputs* is the required number of output nodes
3: # *popSize* is number of organisms in the population
4: # *e* is the number of episodes of evaluation (per organism) to perform
5: # *NEATParams* is a data structure containing the parameters used in evolution, including *AddNodeMutationProb* and *AddLinkMutationProb*
6: $P \leftarrow$ `InitPopulation`(*numInputs, numOutputs, popSize*)
7: **loop**
8:   **for** *evals* $\leftarrow$ 1 to *e* **do**
9:     *ANN, s* $\leftarrow$ *P*[*evals % popsize*], `getStartState()`
10:     **repeat**
11:       **for** $a \in A$ **do**
12:         *Activation*[*a*] $\leftarrow$ *ANN*.`Output`(*ANN.topology, ANN.weights, s, a*)
13:       $a \leftarrow$ `argmax`$_i$(*Activation*[*i*]) # Select action with the highest activation
14:       Execute *a*
15:       Observe *r* and *s*
16:       *ANN.fitness* $\leftarrow$ *ANN.fitness* + *r*
17:     **until** *s* is a terminal state # the episode ends
18:   $P \leftarrow$ `evolvePopulation`(*P, NEATParams*)

There is one output for each available action; the agent takes whichever action has the highest activation.

A candidate policy is evaluated by allowing the corresponding network to control an agent's behavior and observing how much reward it receives. The policy's *fitness* is typically the sum of the rewards the agent accrues while under the network's control. In deterministic domains, each member of the population can be evaluated in a single episode. However, most real-world problems are non-deterministic and hence the reward a policy receives over the course of an episode may have substantial variance. In such domains, it is necessary to evaluate each member of the population for many episodes to accurately estimate policy fitnesses.

Algorithm 4 describes the NEAT algorithm in pseudocode for episodic tasks. In it we assume that the fitness of an organism is its total accumulated reward. This assumption is true in all our experiments, but other fitness functions may also be appropriate. The function `evolvePopulation` takes a number of parameters, as described in the original NEAT paper [Stanley and Miikkulainen (2002)]. Our experiments will make note of parameters when tuned (rather than using the default NEAT settings [Stanley and Miikkulainen (2002)]).

2.3.3 Fitted R-Max

In contrast to Sarsa and NEAT, Fitted R-MAX [Jong and Stone (2007)] is a model-based RL algorithm. While there are many existing model-learning RL algorithms (e.g., [Moore and Atkeson (1993), Kearns and Singh (1998)]), only a handful of other methods (c.f. Bayesian RL methods [Dearden et al. (1999)]) are applicable to non-deterministic tasks with continuous state spaces. Fitted R-MAX is an algorithm that approximates the action-value function for large or infinite state spaces by constructing an abstract MDP over a small (finite) sample of states $X \subset S$. For each sample state $\mathrm{x} \in X$ and action $a \in A$, Fitted R-MAX estimates the dynamics $T(\mathrm{x}, a)$ using all the available data for action a and for states s near x.

Some generalization from nearby states is necessary because we cannot expect the agent to be able to visit x enough times to try every action. As a result of this generalization process, Fitted R-MAX first approximates $T(\mathrm{x}, a)$ as a probability distribution over predicted successor states in S (see Figure 2.8). A value approximation step then approximates this distribution of states in S with a distribution of states in X. The result is a stochastic MDP over a finite state space X, with transition and reward functions derived from observed data. Applying dynamic programming to this MDP yields an action-value function $Q_{model} : X \times A \mapsto \mathbb{R}$ that can be used to approximate the desired action-value function $Q_{task} : S \times A \mapsto \mathbb{R}$. In some benchmark RL tasks, [Jong and Stone (2007)] show empirically that Fitted R-MAX learns policies using less data than many existing model-free algorithms.

In the version of Fitted R-MAX used in this monograph [Jong and Stone (2007)], all experienced instances are recorded. As the model stores more instances, updating the model with additional data becomes slower (although the running time of dynamic programming is bounded because X is bounded). However, a more complete implementation of Fitted R-MAX could include a mechanism to discard instances that were similar enough to existing stored instances that approximate the fitted model, thereby bounding the memory requirements.

Four parameters must be set for Fitted R-MAX to efficiently learn a policy in an MDP. The first, *resolution*, determines the size of the set X (i.e., the number of sample states used for planning). In general, higher resolutions will lead to higher asymptotic performance at the cost of increased computational complexity.

The second, the *model breadth*, helps weight different instances when estimating T. For any state $\mathrm{x} \in X$ and action $a \in A$, Fitted R-MAX estimates $T(\mathrm{x}, a)$ using a probability distribution over observed instances, $\langle s_i, a_i, r_i, s'_i \rangle$, in the data available

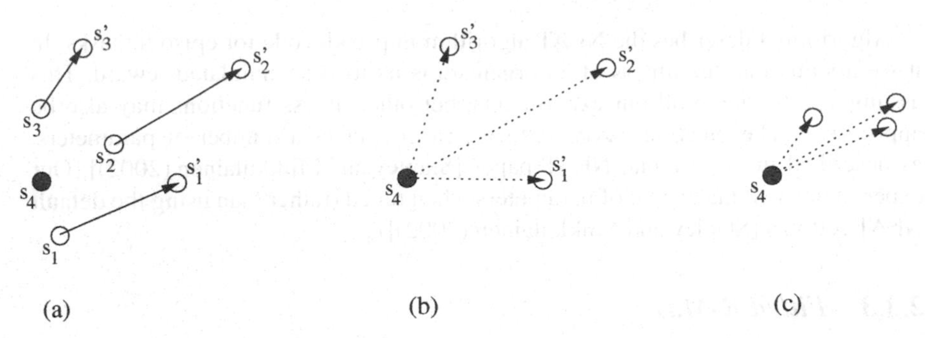

Fig. 2.8 (a) Three observed transitions, all of which execute the same action, can be used to approximate $T(s_4, a)$. (b) One simple way of approximating the next state after executing a is to average the existing next states near s_4. (c) A more appropriate approximation, and one of the key insights of Fitted R-MAX, is to use the *relative* transitions to approximate the effect of an action from a novel state.

for action a. Each instance i is given a weight w_i depending on the Euclidean distance from x to s_i and on the model breadth parameter b, according to the following formula:

$$w_i \propto e^{-\left(\frac{|\mathrm{x}-s_i|}{b}\right)^2}. \tag{2.6}$$

Intuitively, b controls the degree of generalization used to estimate $T(\mathrm{x}, a)$ from nearby data. As the breadth increases, the computational complexity increases, the sample complexity decreases, and the asymptotic policy performance decreases (due to generalization errors).

The third parameter, the *exploration threshold*, is used to determine when enough data exists to accurately estimate some $T(\mathrm{x}, a)$. When an instance is used to approximate $T(\mathrm{x}, a)$, that instance's weight (Equation 2.6) is added to the total weight of the approximation. If the total weight for an approximation does not reach an exploration threshold, an optimistic value (R_{max}) is used because not enough data exists for an accurate approximation. By assigning optimistic values to unknown transitions, the agent is encouraged to explore areas of the space where it has not experienced many transitions and potentially increase the weight of approximations in that region. In our experiments we left this threshold value set at the default, 1.0, from [Jong and Stone (2007)].

The fourth parameter is used to decrease the computational complexity of the dynamic programming step. In theory, all instances that share the action a could be used to help approximate x, where each instance i's contribution is modified by w_i (a Gaussian weighting that exponentially penalizes distance from x). To reduce the computational cost of the algorithm when approximating x, weights for the nearest instances are computed first. Once a single instance's weight fails to increase the cumulative weight by at least some *minimum fraction*, the remaining instances' contributions are ignored as negligible.

Fitted R-MAX is summarized in Algorithm 5. s^{opt} is a dummy state that represents unexplored states (where $V(s^{\mathrm{opt}})$ is set to R_{max}). s^{term} is a dummy absorbing

Algorithm 5. Fitted R-MAX(R_{max}, r, b, *minFraction*, *explorationThreshold*)

```
1: X ← {s^opt, s^term}            # Initialize state sample
2: X.InitializeUniformGrid(r)
3: for all a ∈ A do {             # Initialize experience sample}
4:    D^a ← {⟨s^opt, a, V^max, s^term⟩}
5: loop
6:    s ← initial state           # Begin a trajectory
7:    a ← argmax_a [R(s) + Σ_{x'∈X} P(x'|s,a)V(x')]
8:    repeat
9:       Execute a
10:      Observe r and s'
11:      if s' is terminal then
12:         s' ← s^term
13:      else
14:         a' ← argmax_a [R(s) + Σ_{x'∈X} P(x'|s,a)V(x')]
15:      D^a ← D^a ∪ {⟨s,a,r,s'⟩}        # Update experience sample
16:      Use experience to update φ^X and φ^{S^a}, accounting for minFraction and ex-
         plorationThreshold
17:      Update estimates of R and P based on φ^X and φ^{S^a}
18:      Compute V(x) for x ∈ X via dynamic programming
19:      s ← s'
20:      a ← a'
21:   until s is a terminal state        # the episode ends
```

state that all discovered terminal states get mapped to. D is a data structure that holds all observed instances. ϕ^{S^a} is an averaging object that approximates the effect of action a at state s using nearby sample transitions $d \in D^a$. ϕ^X is an averaging object that approximates the value of each predicted successor state using nearby sample states $x \in X$. The reader is referred to [Jong and Stone (2007)] for detailed descriptions of the update rules (lines 16 and 17).

Chapter 3
Related Work

3.1 Transfer Approaches

Having provided a brief overview of the RL notation used in this monograph, we now enumerate possible approaches for transfer between RL tasks. This section lists attributes of methods used in the TL literature for each of the five dimensions discussed in Section 1.2.2, and summarizes the surveyed works in Table 3.1. The first two groups of methods apply to tasks which have the same state variables and actions. (Section 3.2 discusses the TL methods in the first block, and Section 3.3 discusses the MTL methods in the second block.) Groups three and four consider methods that transfer between tasks with different state variables and actions. (Section 3.4 discusses methods that use a representation that does not change when the underlying MDP changes, while Section 3.5 presents methods that must explicitly account for such changes.) The last group of methods (discussed in Section 3.6) learns a mapping between tasks like those used by methods in the fourth group of methods. Table 3.2 concisely enumerates the possible values for the attributes, as well as providing a key to Table 3.1.

An alternate TL framework may be found in the related work section of [Lazaric (2008)], a recent PhD monograph on TL in RL tasks. Lazaric compares TL methods in terms of the type of benefit (jumpstart, total reward, and asymptotic performance), the allowed differences between source and target (different goal states, different transition functions but the same reward function, and different state and action spaces) and the type of transferred knowledge (experience or structural knowledge). This chapter differs both in the number of approaches considered, the depth of description about each approach, and also uses a different organizational structure. In particular, we specify which of the methods improve which of five TL metrics, we note which of the methods account for source task training time rather than treating it as a sunk cost, and we differentiate methods according to five dimensions above.

In this section the *mountain car* task [Moore (1991), Singh and Sutton (1996)], a standard RL benchmark, will serve as a running example. In mountain car, an under-powered car moves along a curve and attempts to reach a goal state at the

M.E. Taylor: Transfer in Reinforcement Learning Domains, SCI 216, pp. 31–60.
springerlink.com

top of the right "mountain" by selecting between three actions on every timestep: {Forward, Neutral, Backward}, where Forward accelerates the car in the positive x direction and Backward accelerates the car in the negative x direction. The agent's state is described by two state variables: the horizontal position, x, and velocity, $\dot{x}$. The agent receives a reward of -1 on each time step. If the agent reaches the goal state the episode ends and the agent is reset to the start state (often the bottom of the hill, with zero velocity).

3.1.1 Transfer Methods Categorization

3.1.1.1 Allowed Task Differences

TL methods can transfer between MDPs that have different transition functions (denoted by t in Table 3.1), state spaces (s), start states (s_i), goal states (s_f), state variables (v), reward functions (r), and/or action sets (a). For two of the methods, the agent's representation of the world (the *agent space*, describing physical sensors and actuators) remains the same, while the true state variables and actions (the *problem space*, describing the task's state variables and macro-actions) can change (p in Table 3.1). There is also a branch of work that focuses on transfer between tasks which are composed of some number of objects that may change between the source and the target task, such as when learning with RRL (# in Table 3.1). When summarizing the allowed task differences, we will concentrate on the most salient features. For instance, when the source task and target task are allowed to have different state variables and actions, the state space of the two tasks is different because the states are described differently, and the transition function and reward function must also change, but we only indicate "a" and "v."

These differences in the example mountain car task could be exhibited as:

- t: using a more powerful car motor or changing the surface friction of the hill
- s: changing the range of the state variables
- s_i: changing where the car starts each episode
- s_f: changing the goal state of the car
- v: describing the agent's state only by its velocity
- r: rather than a reward of -1 on every step, the reward could be a function of the distance from the goal state
- a: disabling the Neutral action
- p: the agent could describe the state by using extra state variables, such as the velocity on the previous timestep, but the agent only directly measures its current position and velocity
- #: the agent may need to control two cars simultaneously on the hill

3.1.1.2 Source Task Selection

The simplest method for selecting a source task for a given target task is to assume that only a single source task has been learned and that a human has picked it, assuring that the agent should use it for transfer (h in Table 3.1). Some TL algorithms allow the agent to learn multiple source tasks and then use them all for transfer (all). More sophisticated algorithms build a library of seen tasks and use only the most relevant for transfer (lib). Some methods are able to automatically modify a single source task so that the knowledge it gains from the modified task will likely be more useful in the target task (mod). However, none of the existing TL algorithms for RL can guarantee that the source tasks will be useful; a current open question is how to robustly avoid attempting to transfer from an irrelevant task.

3.1.1.3 Transferred Knowledge

The type of knowledge transferred can be primarily characterized by its specificity. Low-level knowledge, such as $\langle$ s, a, r, s′ $\rangle$ instances (I in Table 3.1), an action-value function (Q), a policy (π), a full task model (model), or prior distributions (pri), could all be directly leveraged by the TL algorithm to initialize a learner in the target task. Higher level knowledge, such as what action to use in some situations (A: a subset of the full set of actions), partial policies or options (π_p), rules or advice (rule), important features for learning (fea), proto-value functions (pvf: a type of learned feature), shaping rewards (R), or subtask definitions (sub) may not be directly used by the algorithm to fully define an initial policy, but such information may help guide the agent during learning in the target task.

3.1.1.4 Task Mappings

The majority of TL algorithms in this chapter assume that no explicit task mappings are necessary because the source and target task have the same state variables and actions. In addition to having the same labels, the state variables and actions need to have the same semantic meanings in both tasks. For instance, consider again the mountain car domain. Suppose that the source task had the actions $A = \{$`Forward, Neutral, Backward`$\}$. If the target task had the actions $A = \{$`Right, Neutral, Left`$\}$, a TL method would need some kind of mapping because the actions had different labels (See Sec.5.1). Furthermore, suppose that the target task had the same actions as the source ($A = \{$`Forward, Neutral, Backward`$\}$) but the car was facing the opposite direction, so that `Forward` accelerated the car in the negative x direction and `Backward` accelerated the car in the positive x direction. If the source and target task actions have different semantic meanings, there will also need to be some kind of inter-task mapping to enable transfer.

Methods that do not use a task mapping are marked as "N/A" in Table 3.1. TL methods which aim to transfer between tasks with different state variables or actions typically rely on a task mapping to define how the tasks are related (as defined in Section 3.1.3). Methods that use mappings and assume that they are human-supplied mappings are marked as "sup" in Table 3.1. A few algorithms leverage experience gained in the source task and target task (exp) or a high-level description of the MDPs in order to learn task mappings.

Methods using description-level knowledge differ primarily in what assumptions they make about what will be provided. One method assumes a qualitative understanding of the transition function (T), which would correspond to knowledge like "taking the action Neutral tends to have a positive influence on the velocity in the positive x direction." Two methods assume knowledge of one mapping (M_a: the "action mapping") to learn a second mapping (the "state variable mapping" in Section 3.1.3). Three methods assume that the state variables are "grouped" together to describe objects (sv_g). An example of the state variable grouping can be demonstrated in a mountain car task with multiple cars: if the agent knew which position state variables referred to the same car as certain velocity state variables, it would know something about the grouping of state variables. These different assumptions are discussed in detail in Section 3.6.

3.1.1.5 Allowed Learners

The type of knowledge transferred directly affects the type of learner that is applicable (as discussed in Section 1.3.1). For instance, a TL method that transfers an action-value function would likely require that the target task agent use a temporal difference method to exploit the transferred knowledge. The majority of methods in the literature use a standard form of temporal difference learning (TD in Table 3.1), such as Sarsa. Other methods include Bayesian learning (B), hierarchical approaches (H), model-based learning (MB), direct policy search (PS), and relational reinforcement learning (RRL). Some TL methods focus on batch learning (Batch), rather than on-line learning. Two methods use *case based reasoning* (CBR) [Aamodt and Plaza (1994)] to help match previously learned instances with new instances, and one uses linear programming (LP) to calculate a value function from a given model (as part of a dynamic programming routine).

3.1.2 Multi-Task Learning

Closely related to TL algorithms, and discussed in Section 3.3, are *multi-task learning* (MTL) algorithms. The primary distinction between MTL and TL is that multi-task learning methods assume all problems experienced by the agent are drawn from the same distribution, while TL methods may allow for arbitrary source and target tasks. For example, a MTL task could be to learn a series of mountain car tasks, each of which had a transition function that was drawn from a fixed distribution of functions that specified a range of surface frictions. Because of this assumption, MTL methods generally do not need task mappings (dimension III in Section 1.2.2). MTL algorithms may be used to transfer knowledge between learners, similar to TL algorithms, or they can attempt to learn how to act on the entire class of problems.

In MTL, problems can be learned simultaneously, which is particularly appropriate for multiagent settings [Stone and Veloso (2000)], as each agent could be considered as learning a slightly different task. In transfer learning settings, the assumption is that the source task is first learned and then the target task is learned.

Table 3.1 This table lists all the TL methods discussed in this chapter and classifies each in terms of the five transfer dimensions (the key for abbreviations is in Table 3.2). Two entries, marked with a ⋆, are repeated due to multiple contributions. Metrics that account for source task learning time, rather than ignoring it, are marked with a †.

Citation	Allowed Task Differences	Source Task Selection	Task Mappings	Transferred Knowledge	Allowed Learners	TL Metrics
Same state variables and actions: Section 3.2						
[Selfridge et al. (1985)]	t	h	N/A	Q	TD	tt†
[Asada et al. (1994)]	s_i	h	N/A	Q	TD	tt
[Singh (1992)]	r	all	N/A	Q	TD	ap, tr
[Atkeson and Santamaria (1997)]	r	all	N/A	model	MB	ap, j, tr
[Asadi and Huber (2007)]	r	h	N/A	π_p	H	tt
[Andre and Russell (2002)]	r, s	h	N/A	π_p	H	tr
[Ravindran and Barto (2003b)]	s, t	h	N/A	π_p	TD	tr
[Ferguson and Mahadevan (2006)]	r, s	h	N/A	pvf	Batch	tt
[Sherstov and Stone (2005)]	s_f, t	mod	N/A	A	TD	tr
[Madden and Howley (2004)]	s, t	all	N/A	rule	TD	tt, tr
[Lazaric (2008)]	s, t	lib	N/A	I	Batch	j, tr
Multi-Task learning: Section 3.3						
[Mehta et al. (2008)]	r	lib	N/A	π_p	H	tr
[Perkins and Precup (1999)]	t	all	N/A	π_p	TD	tt
[Foster and Dayan (2004)]	s_f	all	N/A	sub	TD, H	j, tr
[Fernandez and Veloso (2006)]	s_i, s_f	lib	N/A	π	TD	tr
[Tanaka and Yamamura (2003)]	t	all	N/A	Q	TD	j, tr
[Sunmola and Wyatt (2006)]	t	all	N/A	pri	B	j, tr
[Wilson et al. (2007)]	r, s_f	all	N/A	pri	B	j, tr
[Walsh et al. (2006)]	r, s	all	N/A	fea	any	tt
[Lazaric (2008)]⋆	r	all	N/A	fea	Batch	ap, tr
Different state variables and actions – no explicit task mappings: Section 3.4						
[Konidaris and Barto (2006)]	p	h	N/A	R	TD	j, tr
[Konidaris and Barto (2007)]	p	h	N/A	π_p	TD	j, tr
[Banerjee and Stone (2007)]	a, v	h	N/A	fea	TD	ap, j, tr
[Guestrin et al. (2003)]	#	h	N/A	Q	LP	j
[Croonenborghs et al. (2007)]	#	h	N/A	π_p	RRL	ap, j, tr
[Ramon et al. (2007)]	#	h	N/A	Q	RRL	ap, j, tt†, tr
[Sharma et al. (2007)]	#	h	N/A	Q	TD, CBR	j, tr
Different state variables and actions – inter-task mappings used: Section 3.5						
[Taylor et al. (2007a)]	a, v	h	sup	Q	TD	tt†
[Taylor et al. (2007b)]	a, v	h	sup	π	PS	tt†
[Taylor et al. (2008b)]	a, v	h	sup	I	MB	ap, tr
[Torrey et al. (2005)] [Torrey et al. (2006)]	a, r, v	h	sup	rule	TD	j, tr
[Torrey et al. (2007)]	a, r, v	h	sup	π_p	TD	j, tr
[Taylor and Stone (2007a)]	a, r, v	h	sup	rule	any/TD	j, tt†, tr
[Taylor and Stone (2007b)]	a, v	h	sup	I	TD, PS	j, tt†, tr
Learning inter-task mappings: Section 3.6						
[Kuhlmann and Stone (2007)]	a, v	h	T	Q	TD	j, tr
[Liu and Stone (2006)]	a, v	h	T	N/A	all	N/A
[Soni and Singh (2006)]	a, v	h	M_a, sv_g, exp	N/A	all	ap, j, tr
[Talvitie and Singh (2007)]	a, v	h	M_a, sv_g, exp	N/A	all	j
[Taylor et al. (2007b)]⋆	a, v	h	sv_g, exp	N/A	all	tt†
[Taylor et al. (2008d)]	a, v	h	exp	N/A	all	j, tr

Table 3.2 This key provides a reference to the abbreviations in Table 3.1

Allowed Task Differences	
a	action set may differ
p	problem space may differ (agent space must be identical)
r	reward function may differ
s_i	the start state may change
s_f	goal state may move
t	transition function may differ
v	state variables may differ
#	number of objects in state may differ

Source Task Selection	
all	all previously seen tasks are used
h	one source task is used (human selected)
lib	tasks are organized into a library and one or more may be used
mod	a human provides a source task that the agent automatically modifies

Task Mappings	
exp	agent learns the mappings from experience
M_a	the method must be provided with an action mapping (learns state variable mapping)
N/A	no mapping is used
sup	a human supplies the task mappings
sv_g	method is provided groupings of state variables
T	higher-level knowledge is provided about transfer functions to learn mapping

Transferred Knowledge	
A	an action set
fea	task features
I	experience instances
model	task model
π	policies
π_p	partial policies (e.g., options)
pri	distribution priors
pvf	proto-value function
Q	action-value function
R	shaping reward
rule	rules or advice
sub	subtask definitions

Allowed Learners	
B	Bayesian learner
Batch	batch learner
CBR	case based reasoning
H	hierarchical value-function learner
LP	linear programming
MB	model based learner
PS	policy search learner
RRL	relational reinforcement learning
TD	temporal difference learner

TL Metrics	
ap	asymptotic performance increased
j	jumpstart demonstrated
tr	total reward increased
tt	task learning time reduced

[Sutton et al. (2007)] motivate this approach to transfer by suggesting that a single large task may be most appropriately tackled as a sequential series of subtasks. If the learner can track which subtask it is currently in, it may be able to transfer knowledge between the different subtasks, which are all presumably related because they are part of the same overall task. Such a setting may provide a well-grounded way of selecting a distribution of tasks to train over, either in the context of transfer or for multi-task learning. Note also that the additional assumptions in an MTL setting may be leveraged to allow a more rigorous theoretical analysis than in TL (c.f., [Kalmár and Szepesvári (1999)]).

3.1.3 Inter-Task Mappings

Transfer methods that assume the source and target tasks use the same state variables and actions, as is the case in MTL, typically do not need an explicit mapping between tasks. In order to enable TL methods to transfer between tasks that do have such differences, the agent must know how the tasks are related. This section provides a brief overview of *inter-task mappings*, discussed in more detail in Section 5.1.1. Task mappings like these are used by transfer methods discussed in Section 3.5.

To transfer effectively, when an agent is presented with a target task that has a set of actions (A'), it must know how those actions are related to the action set in the source task (A). (For the sake of exposition we focus on actions, but an analogous argument holds for state variables.) If the TL method knows that the two action sets are identical, no action mapping is necessary. However, if this is not the case, the agent needs to be told, or learn, how the two tasks are related. For instance, if the agent learns to act in a source task with the actions `Forward` and `Backward`, but the target task uses the actions `Right` and `Left`, the correspondence between these action sets may not be obvious. Even if the action labels were the same, if the actions had different semantic meanings, the default correspondence may be incorrect. Furthermore, if the cardinality of A and A' are not equal, there are actions without exact equivalences.

One option is to define an *action mapping* (χ_A) such that actions in the two tasks are mapped so that their effects are "similar," where similarity depends on the transfer and reward functions in the two MDPs.[1] Figure 3.1 depicts an action mapping as well as a *state-variable mapping* (χ_X) between two tasks. A second option is to define a *partial mapping* [Taylor et al. (2007b)], such that any novel actions in the target task are ignored. Consider adding an action in a mountain car target task, `pull hand brake`, which did not have an analog in the source task. The partial mapping could map `Forward` to `Forward`, and `Backward` to `Backward`, but not map `pull hand brake` to any source task action. Because inter-task mappings are not functions, they are typically assumed to be easily invertible (i.e., mapping

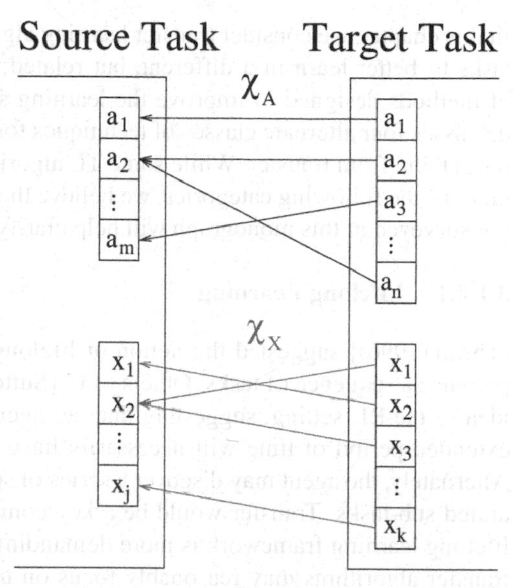

Fig. 3.1 χ_A and χ_X are independent mappings that describe similarities between two MDPs. These mappings describe how actions in the target task are similar to actions in the source task and how state variables in the target task are similar to state variables in the source task, respectively.

[1] An inter-task mapping often maps multiple entities in the target task to single entities in the source task because the target task is more complex than the source, but the mappings may be one-to-many, one-to-one, or many-to-many.

source task actions into target task actions, rather than target task actions to source task actions).

It is possible that mappings between states, rather than between state variables, could be used for transfer, although no work has currently explored this formulation. Another possible extension is to link the mappings rather than making them independent. For instance, the action mapping could depend on the state that the agent is in, or the state variable mapping could depend on the action selected. Though these extensions may be necessary based on the demands of particular MDPs, current methods have functioned well in a variety of tasks without such enhancements.

For a given pair of tasks, there could be many ways to formulate inter-task mappings. Much of the current TL work assumes that a human has provided a (correct) mapping to the learner. Work that attempts to learn a mapping that can be effectively used for transfer is discussed in Section 3.6.

3.1.4 Related Paradigms

In this chapter, we consider transfer learning algorithms that use one or more source tasks to better learn in a different, but related, target task. There is a wide range of methods designed to improve the learning speed of RL methods. This section discusses four alternate classes of techniques for speeding up learning and differentiates them from transfer. While some TL algorithms may reasonably fit into one or more of the following categories, we believe that enumerating the types of methods *not* surveyed in this monograph will help clarify our subject of interest.

3.1.4.1 Lifelong Learning

[Thrun (1996)] suggested the notion of lifelong learning where an agent may experience a sequence of tasks. Others (c.f., [Sutton et al. (2007)]) later extended this idea to the RL setting, suggesting than an agent interacting with the world for an extended period of time will necessarily have to perform in a sequence of tasks. Alternately, the agent may discover a series of spatially, rather than temporally, separated sub-tasks. Transfer would be a key component of any such system, but the lifelong learning framework is more demanding than that of transfer. For instance, transfer algorithms may reasonably focus on transfer between a single pair of related tasks, rather than attempting to account for any future task that an agent could encounter.

3.1.4.2 Imitation Learning

The primary motivations for imitation methods are to allow agents to learn by watching another agent with similar abilities [Price and Boutilier (2003), Syed and Schapier (2007)] or a human [Abbeel and Ng (2005), Kolter et al. (2008)] perform a task. Such algorithms attempt to learn a policy by observing an outside actor, potentially improving upon the inferred policy. In contrast, our definition of transfer learning focuses on agents successfully reusing internal knowledge on novel problems.

3.1.4.3 Human Advice

There is a growing body of work integrating human advice into RL learners. For instance, a human may provide action suggestions to the agent (c.f., [Maclin and Shavlik (1996), Maclin et al. (2005)]) or guide the agent through on-line feedback (c.f., [Knox and Stone (2008)]). Leveraging humans' background and task-specific knowledge can significantly improve agents' learning ability, but it relies on a human being tightly integrated into the learning loop. This monograph instead concentrates on transfer methods in which a human is not continuously available and agents must learn autonomously.

3.1.4.4 Shaping

Reward shaping [Colombetti and Dorigo (1993), Mataric (1994)] in an RL context typically refers to allowing agent to train on an artificial reward signal rather than R. For instance, in the mountain car task, the agent could be given a higher reward as it gets closer to the goal state, rather than receiving -1 at every state except the goal. However, if the human can compute such a reward, s/he would probably already know the goal location, knowledge that the agent typically does not have. Additionally, the constructed reward function must be a potential function. If it is not, the optimal policy for the new MDP could be different from that of the original [Ng et al. (1999)]. A second definition of shaping follows Skinner's research [Skinner (1953)] where the reward function is modified over time in order to direct the behavior of the learner. This method, as well as the approach of using a static artificial reward, is a way of injecting human knowledge into the task definition to improve learning efficacy.

[Erez and Smart (2008)] have argued for a third definition of shaping as any supervised, iterative, process to assist learning. This includes modifying the dynamics of the task over time, modifying the internal learning parameters over time, increasing the actions available to the agent, and extending the agent's policy time horizon (e.g., as done in value iteration). All of these methods rely on a human to intelligently assist the agent in its learning task and may leverage transfer-like methods to successfully reuse knowledge between slightly different tasks. When discussing transfer, we will emphasize how knowledge is successfully reused rather than how to best modify tasks to achieve the desired agent behavior improve agent learning performance.

3.2 Transfer Methods for Fixed State Variables and Actions

To begin our survey of TL methods, we examine the first group of methods in Table 3.1, reproduced in Table 3.3. These techniques may be used for transfer when

Table 3.3 This table reproduces the first group of methods from Table 3.1

Citation	Allowed Task Differences	Source Task Selection	Task Mappings	Transferred Knowledge	Allowed Learners	TL Metrics
Same state variables and actions: Section 3.2						
[Selfridge et al. (1985)]	t	h	N/A	Q	TD	tt[†]
[Asada et al. (1994)]	s_i	h	N/A	Q	TD	tt
[Singh (1992)]	r	all	N/A	Q	TD	ap, tr
[Atkeson and Santamaria (1997)]	r	all	N/A	model	MB	ap, j, tr
[Asadi and Huber (2007)]	r	h	N/A	π_p	H	tt
[Andre and Russell (2002)]	r, s	h	N/A	π_p	H	tr
[Ravindran and Barto (2003b)]	s, t	h	N/A	π_p	TD	tr
[Ferguson and Mahadevan (2006)]	r, s	h	N/A	pvf	Batch	tt
[Sherstov and Stone (2005)]	s_f, t	mod	N/A	A	TD	tr
[Madden and Howley (2004)]	s, t	all	N/A	rule	TD	tt, tr
[Lazaric (2008)]	s, t	lib	N/A	I	Batch	j, tr

the source and target tasks use the same state variables and when agents in both tasks have the same set of actions (see Figure 3.2).

In one of the earliest TL works for RL, [Selfridge et al. (1985)] demonstrated that it was faster to learn to balance a pole on a cart by changing the task's transition function, T, over time. The learner was first trained on a long and light pole. Once it successfully learned to balance the pole the task was made harder: the pole was shortened and made heavier. The total time spent training on a sequence of tasks and reusing the learned function approximator was faster than training on the hardest task directly.

Similarly, the idea of *learning from easy missions* [Asada et al. (1994)] also relies on a human constructing a set of tasks for the learner. In this work, the task (for example, a maze) is made incrementally harder not by changing the dynamics of the task, but by moving the agent's initial state, $s_{initial}$, further and further from the goal state. The agent incrementally learns how to navigate to the exit faster than if it had tried to learn how to navigate the full maze directly. This method relies on having a known goal state from which a human can construct a series of source tasks of increasing difficulty.

[Selfridge et al. (1985)] and [Asada et al. (1994)] provide useful methods for improving learning, which follow from Skinner's animal training work. While they require a human to be in the loop, and to understand the task well enough to provide the appropriate guidance to the learner, these methods are relatively easy ways to leverage human knowledge. Additionally, they may be combined with many of the transfer methods that follow.

Rather than change a task over time, one could consider breaking down a task into a series of smaller tasks. This approach can be considered a type of transfer in that a single large target task can be treated as a series of simpler source tasks. [Singh (1992)] uses a technique he labels *compositional learning* to discover how to separate temporally sequential subtasks in a monolithic task. Each subtask has distinct beginning and termination conditions, and each subtask will be significantly easier to learn in isolation than in the context of the full task. Only the reward function, *R*, is allowed to change between the different subtasks and none of the other MDP components may vary, but the total reward can be increased. If subtasks in a problem are recognizable by state features, such subtasks may be automatically identified via vision algorithms [Drummond (2002)]. Again, breaking a task into smaller subtasks can improve both the total reward and the asymptotic performance. This particular method is only directly applicable to tasks in which features clearly define subtasks due to limitations in the vision algorithm used. For instance, in a 2D navigation task each room may be a subtask and the steep value function gradient between impassable walls is easily identifiable. However, if the value function gradient is not distinct between different subtasks, or the subtask regions of state space are not polygonal, the algorithm will likely fail to automatically identify subtasks.

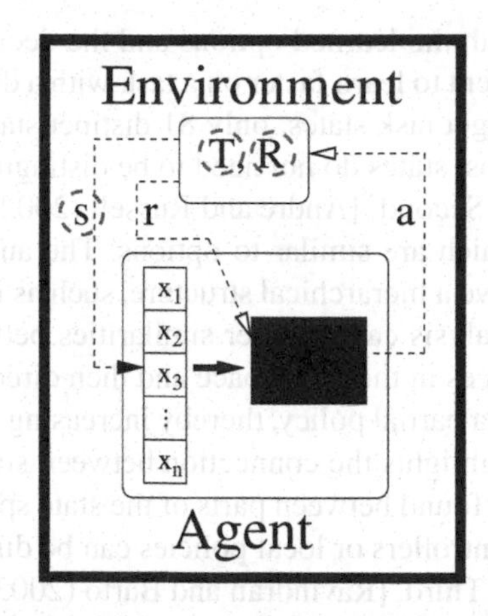

Fig. 3.2 Methods in Section 3.2 are able to transfer between tasks that have different state spaces, different transition functions, and different reward functions, but only if the source and target tasks have the same actions and state variables. Dashed circles indicate the MDP components which may differ between the source task and target task.

In [Atkeson and Santamaria (1997)], transfer between tasks in which only the reward function can differ are again considered. Their method successfully transfers a locally weighted regression model of the transition function, which is learned in a source task, by directly applying it to a target task. Because their model enables planning over the transition function and does not account for the reward function, they show significant improvement to the jumpstart and total reward, as well as the asymptotic performance.

The next three methods transfer partial policies, or options, between different tasks. First, [Asadi and Huber (2007)] have the agent identify states that "locally form a significantly stronger 'attractor' for state space trajectories" as subgoals in the source task (i.e., a doorway between rooms that is visited relatively often compared to other parts of the state space). The agent then learns options to reach these subgoals via a learned action-value function, termed the *decision-level* model. A second action-value function, the *evaluation-level* model, includes all actions and the full state space. The agent selects actions by only considering the decision-level model but uses discrepancies between the two models to automatically increase the complexity of the decision-level model as needed. The model is represented as a *Hierarchical Bounded Parameter SMDP*, constructed so that the performance of an optimal policy in the simplified model will be within some fixed bound of the performance of the optimal policy on the initial model. Experiments show that transferring both the learned options and the decision-level representation allow the target task agent to learn faster on a task with a different reward function. In the roughly 20,000 target task states, only 81 distinct states are needed in the decision-level model, as most states do not need to be distinguished when selecting from learned options.

Second, [Andre and Russell (2002)] transfer learned subroutines between tasks, which are similar to options. The authors assume that the source and target tasks have a hierarchical structure, such as in the *taxi domain* [Dietterich (2000)]. On-line analysis can uncover similarities between two tasks if there are only small differences in the state space and then directly copy over the subroutine, which functions as a partial policy, thereby increasing the total reward in the target task. This method highlights the connection between state abstraction and transfer; if similarities can be found between parts of the state space in the two tasks, it is likely that good local controllers or local policies can be directly transferred.

Third, [Ravindran and Barto (2003b)] learn *relativized options* in a small, human selected source task. When learning in the target task, the agent is provided these options and a set of possible transformations it could apply to them so that they were relevant in the target task. For instance, if the source task were a small grid navigation task, the target task could be a large grid composed of rooms with similar shape to the source task and the transformations could be rotation and reflection operators. The agent uses experience in the target and Bayesian parameter estimation to select which transformations to use so that the target task's total reward is increased. Learning time in the source task is ignored, but is assumed to be small compared to the target task learning time.

Next, [Ferguson and Mahadevan (2006)] take a unique approach that transfers information about the underlying task structure. *Proto-value functions* (PVFs) [Mahadevan and Maggioni (2007)] specify an ortho-normal set of basis functions, without regard to R, which can be used to learn an action-value function. After PVFs are learned in a small source task, they can be transferred to another discrete MDP that has a different goal or small changes to the state space. The target task can be learned faster and achieve higher total reward with the transferred PVFs than without. Additionally, the PVF can be scaled to larger tasks. For example, the target maze could have twice the width and height of the source maze: R, S, and T are all scaled by the same factor. In all cases only the target task time is counted.

The goal of learning PVFs is potentially very useful for RL in general and TL in particular. It makes intuitive sense that high-level information about how to best learn in a domain, such as appropriate features to reason over, may transfer well across tasks. There are few examples of meta-learners where TL algorithms learn high level knowledge to assist the agent in learning, rather than lower-level knowledge about how to act. However, we believe that there is ample room for such methods, including methods to learn other domain-specific learning parameters, such as learning rates, function approximator representations, and so on.

Instead of biasing the target task agent's learning representation by transferring a set of basis functions, [Sherstov and Stone (2005)] consider how to bias an agent by transferring an appropriate action set. If tasks have large action sets, all actions could be considered when learning each task, but learning would be much faster if only a subset of the actions needed to be evaluated. If a reduced action set is selected such that using it could produce near-optimal behavior, learning would be much faster with very little loss in final performance. The standard MDP formalism is modified so that the agent reasons about *outcomes* and *classes*. Informally, rather than reasoning over the probability of reaching a given state after an action, the learner reasons over the actions' effect, or outcome. States are grouped together in classes such that the probability of a given outcome from a given action will be the same for any state in a class. The authors then use their formalism to bound the value lost by using their abstraction of the MDP. If the source and target are very similar, the source task can be learned with the full action set, the optimal action set can be found from the learned Q-values, and learning the target with this smaller action set can speed up learning in the target task. The authors also introduce *random task perturbation* (RTP) which creates a *series* of source tasks from a single source task, thereby producing an action set which will perform well in target tasks that are less similar to the source task. Transfer with and without RTP is experimentally compared to learning without transfer. While direct action transfer can perform worse than learning without transfer, RTP was able to handle misleading source task experience so that performance was improved relative to no transfer in all target tasks and performance using the transferred actions approaches that of the optimal target task action set. Performance was judged by the total reward accumulated in the target task. [Leffler et al. (2007)] extends the work of Sherstov and Stone by applying the outcome/class framework to learn a *single* task significantly faster, and provides empirical evidence of correctness in both simulated and physical domains.

The idea of RTP is not only unique in this survey, but it is also potentially a very useful idea for transfer in general. While a number of TL methods are able to learn from a set of source tasks, no others attempt to automatically generate these source tasks. If the goal of an agent is perform as well as possible in a novel target task, it makes sense that the agent would try to train on many source tasks, even if they are artificial. How to best generate such source tasks so that they are most likely to be useful for an arbitrary target task in the same domain is an important area of open research.

Similar to previously discussed work [Selfridge et al. (1985), Asada et al. (1994)], *Progressive RL* [Madden and Howley (2004)] is a method for transferring between a progression of tasks of increasing difficulty, but is limited to discrete MDPs. After learning a source task, the agent performs *introspection* where a symbolic learner extracts rules for acting based on learned Q-values from all previously learned tasks. The RL algorithm and introspection use different state features. Thus the two learning mechanisms learn in different state spaces, where the state features for the symbolic learner are higher-level and contain information otherwise hidden from the agent. When the agent acts in a novel task, the first time it reaches a novel state it initialize the Q-values of that state so that the action suggested by the learned rule is preferred. Progressive RL allows agents to learn information in a set of tasks and then abstract the knowledge to a higher-level representation, allowing the agent to achieve higher total reward and reach the goal state for the first time faster. Time spent in the source task(s) is not counted.

Finally, [Lazaric (2008)] demonstrates that source task *instances* can be usefully transferred between tasks. After learning one or more source tasks, some experience is gathered in the target task, which may have a different state space or transition function. Saved instances (that is, observed $\langle s,a,r,s' \rangle$ tuples) are compared to instances from the target task. Instances from the source tasks that are most similar, as judged by their distance and alignment with target task data, are transferred. A batch learning algorithm then uses both source instances and target instances to achieve a higher reward and a jumpstart. *Region transfer* takes the idea one step further by looking at similarity with the target task per-sample, rather than per task. Thus, if source tasks have different regions of the state space which are more similar to the target, only those most similar regions can be transferred. In these experiments, time spent training in the target task is not counted towards the TL algorithm.

Region transfer is the only method surveyed which explicitly reasons about task similarity in *different parts* of the state space, and then selects source task(s) to transfer from. In domains where target tasks have regions of the state space that are similar to one or more source tasks, and other areas which are similar to other source tasks (or are similar to no source tasks), region transfer may provide significant performance improvements. As such, this method provides a unique approach to measuring, and exploiting, task similarity on-line. It is likely that this approach will inform future transfer methods, and is one possible way of accomplishing step # 1 in Section 1.2: Given a target task, select an appropriate source task from which to transfer, if one exists.

Taken together, these TL methods show that it is possible to efficiently transfer many different types of information between tasks with a variety of differences. It is worth re-emphasizing that many TL methods may be combined with other speedup methods, such as reward shaping, or with other transfer methods. For instance, when transferring between maze tasks, basis functions could be learned [Ferguson and Mahadevan (2006)] in the source task, a set of actions to transfer could be selected after training on a set of additional generated source tasks [Sherstov and Stone (2005)], and then parts of different source tasks could be leveraged to learn a target task [Lazaric (2008)]. A second example would be to start with a simple source task and change it over time by modifying the transition function [Selfridge et al. (1985)] and start state [Asada et al. (1994)], while learning options [Ravindran and Barto (2003b)], until a difficult target task is learned. By examining how the source and target task differ and what base learning method is used, RL practitioners may select one or more TL method to apply to their domain of interest. However, in the absence of theoretical guarantees of transfer efficacy, any TL method has the potential to be harmful, as discussed further in Section 9.3.1.

3.3 Multi-Task Learning Methods

This section discusses scenarios where the source tasks and target task have the same state variables and actions. However, these methods (see Table 3.4, reproduced from Table 3.1) are explicitly MTL, and all methods in this section are designed to use multiple source tasks (see Figure 3.3). Some methods leverage all experienced

Table 3.4 This table reproduces the group of MTL methods from Table 3.1

Citation	Allowed Task Differences	Source Task Selection	Task Mappings	Transferred Knowledge	Allowed Learners	TL Metrics
Multi-Task learning: Section 3.3						
[Mehta et al. (2008)]	r	lib	N/A	π_p	H	tr
[Perkins and Precup (1999)]	t	all	N/A	π_p	TD	tt
[Foster and Dayan (2004)]	s_f	all	N/A	sub	TD, H	j, tr
[Fernandez and Veloso (2006)]	s_i, s_f	lib	N/A	π	TD	tr
[Tanaka and Yamamura (2003)]	t	all	N/A	Q	TD	j, tr
[Sunmola and Wyatt (2006)]	t	all	N/A	pri	B	j, tr
[Wilson et al. (2007)]	r, s_f	all	N/A	pri	B	j, tr
[Walsh et al. (2006)]	r, s	all	N/A	fea	any	tt
[Lazaric (2008)]	r	all	N/A	fea	Batch	ap, tr

source tasks when learning a novel target task and others are able to choose a subset of previously experienced tasks. Which approach is most appropriate depends on the assumptions about the task distribution: if tasks are expected to be similar enough that all past experience is useful, there is no need to select a subset. On the other hand, if the distribution of tasks is multi-modal, it is likely that transferring from all tasks is sub-optimal. None of the methods account for time spent learning in the source task(s) as the primary concern is effective learning on the next task chosen at random from an unknown (but fixed) distribution of MDPs.

Variable-reward hierarchical reinforcement learning [Mehta et al. (2008)] assumes that the learner will train on a sequence of tasks which are identical except for different *reward weights*. The reward weights define how much reward is assigned via a linear combination of *reward features*. The authors provide the reward features to the agent for a given set of tasks. For instance, in a real-time strategy

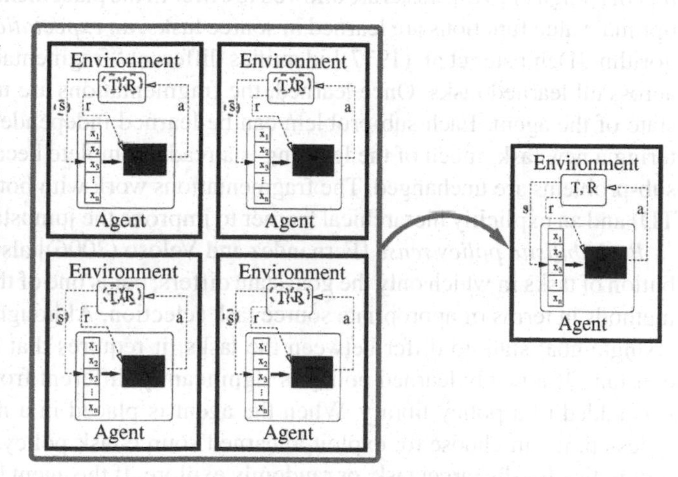

Fig. 3.3 Multi-task learning methods assume tasks are chosen from a fixed distribution, use one or more source tasks to help learn the current task, and assume that all the tasks have the same actions and state variables. Dashed circles indicate the MDP components which may differ between tasks.

domain different tasks could change the reward features, such as the benefit from collecting units of gold or from damaging the enemy. However, it is unclear how many domains of interest have reward features, which are provided to the agent at the start of each task. Using a hierarchical RL method, subtask policies are learned. When a novel target task is encountered, the agent sets the initial policy to that of the most similar source task, as determined by the dot product with previously observed reward weight vectors. The agent then uses an ε-greedy action selection method at each level of the task hierarchy to decide whether to use the best known sub-task policy or explore. Some sub-tasks, such as navigation, will never need to be relearned for different tasks because they are unaffected by the reward weights, but any suboptimal sub-task policies will be improved. As the agent experiences more tasks, the total reward in each new target task increases, relative to learning the task without transfer.

A different problem formulation is posed by [Perkins and Precup (1999)] where the transition function, T, may change after reaching the goal. Upon reaching the goal, the agent is returned to the start state and is not told if, or how, the transition function has changed, but it knows that T is drawn randomly from some fixed distribution. The agent is provided a set of hand-coded options which assist in learning on this set of tasks. Over time, the agent learns an accurate action-value function over these options. Thus, a single action-value function is learned over a set of tasks, allowing the agent to more quickly reach the goal on tasks with novel transition functions.

Instead of transferring options, [Foster and Dayan (2004)] aim to identify sub-tasks in a source task and use this information in a target task, a motivation similar to that of [Singh (1992)]. Tasks are allowed to differ in the placement of the goal state. As optimal value functions are learned in source tasks, an *expectation-maximization* algorithm [Dempster et al. (1977)] identifies different "fragmentations," or sub-tasks, across all learned tasks. Once learned, the fragmentations are used to augment the state of the agent. Each sub-problem can be learned independently; when encountering a new task, much of the learning is already complete because the majority of sub-problems are unchanged. The fragmentations work with both a flat learner (i.e., TD) and an explicitly hierarchical learner to improve the jumpstart and total reward.

Probabilistic policy reuse [Fernandez and Veloso (2006)] also considers a distribution of tasks in which only the goal state differs, but is one of the most robust MTL methods in terms of appropriate source task selection. Although the method allows a single goal state to differ between the tasks, it requires that S, A, and T remain constant. If a newly learned policy is significantly different from existing policies, it is added to a policy library. When the agent is placed in a novel task, on every timestep, it can choose to: exploit a learned source task policy, exploit the current best policy for the target task, or randomly explore. If the agent has multiple learned policies in its library, it probabilistically selects between policies so that over time more useful policies will be selected more often. While this method allows for probabilistic mixing of the policies, it may be possible to treat the past policies as options which can be executed until some termination condition is met, similar to a number of previously discussed methods. By comparing the relative benefits of mixing past policies and treating them as options, it may be possible to better understand when each of the two approaches is most useful.

The idea of constructing an explicit policy library is likely to be useful in future TL research, particularly for agents that train on a number of source tasks that have large qualitative differences (and thus very different learned behaviors). Although other methods also separately record information from multiple source tasks (c.f., [Mehta et al. (2008), Lazaric (2008)]), Fernandez and Veloso explicitly reason about the library. In addition to reasoning over the amount of information stored, as a function of number and type of source tasks, it will be useful to understand how many target task samples are needed to select the most useful source task(s).

Unlike probabilistic policy reuse, which selectively transfers information from a single source task, [Tanaka and Yamamura (2003)] gather statistics about *all* previous tasks and use this amalgamated knowledge to learn novel tasks faster. Specifically, the learner keeps track of the average and the deviation of the action value for each (s,a) pair observed in all tasks. When the agent encounters a new task, it initializes the action-value function so that every (s,a) pair is set to the current average for that pair, which provides a benefit relative to uninformed initialization. As the agent learns the target task with Q-learning and prioritized sweeping[2], the agent uses the standard deviation of states' Q-values to set priorities on TD backups. If the current Q-value is far from the average for that (s,a) pair, its value should be adjusted more quickly, since it is likely incorrect (and thus should be corrected before affecting other Q-values). Additionally, another term accounting for the variance within individual trials is added to the priority; Q-values that fluctuate often within a particular trial are likely wrong. Experiments show that this method, when applied to sets of discrete tasks with different transition functions, can provide significant improvement to jumpstart and total reward.

The next two methods consider how priors can be effectively learned by a Bayesian MTL agent. First, [Sunmola and Wyatt (2006)] introduce two methods that use instances from source tasks to set priors in a Bayesian learner. Both methods constrain the probabilities of the target task's transition function by using previous instances as a type of prior. The first method uses the working prior to generate possible models which are then tested against data in the target task. The second method uses a probability perturbation method in conjunction with observed data to improve models generated by the prior. Initial experiments show that the jumpstart and total reward can be improved if the agent has an accurate estimation of the prior distributions of the class from which the target is drawn. Second, [Wilson et al. (2007)] consider learning in a hierarchical Bayesian RL setting. Setting the prior for Bayesian models is often difficult, but in this work the prior may be transferred from previously learned tasks, significantly increasing the learning rate. Additionally, the algorithm can handle "classes" of MDPs, which have similar model parameters, and

[2] Prioritized sweeping [Moore and Atkeson (1993)] is an RL method that orders adjustments to the value function based on their "urgency," which can lead to faster convergence than when updating the value function in the order of visited states.

then recognize when a novel class of MDP is introduced. The novel class may then be added to the hierarchy and a distinct prior may be learned, rather than forcing the MDP to fit into an existing class. The location of the goal state and the parameterized reward function may differ between the tasks. Learning on subsequent tasks shows a clear performance improvement in total reward, and some improvement in jumpstart.

While Bayesian methods have been shown to be successful when transferring between classification tasks [Roy and Kaelbling (2007)], and in non-transfer RL [Dearden et al. (1999)], only the two methods above use it in RL transfer. The learner's bias is important in all machine learning settings. However, Bayesian learning makes such bias explicit. Being able to set the bias through transfer from similar tasks may prove to be a very useful heuristic—we hope that additional transfer methods will be developed to initialize Bayesian learners from past tasks.

[Walsh et al. (2006)] observe that "deciding what knowledge to transfer between environments can be construed as determining the correct state abstraction scheme for a set of source [tasks] and then applying this compaction to a target [task]." Their suggested framework solves a set of MDPs, builds abstractions from the solutions, extracts relevant features, and then applies the feature-based abstraction function to a novel target task. A simple experiment utilizing tasks with different state spaces and reward functions shows that the time to learn a target task is decreased by using MTL. Building upon their five defined types of state abstractions (as defined in [Li et al. (2006)]), they give theoretical results showing that when the number of source tasks is large (relative to the differences between the different tasks), four of the five types of abstractions are guaranteed to produce the optimal policy in a target task using Q-learning.

Similar to [Walsh et al. (2006)], [Lazaric (2008)] also discovers features to transfer. Rather than learning tasks sequentially, as in all the papers above, one could consider learning different tasks in parallel and using the shared information to learn the tasks better than if each were learned in isolation. Specifically, [Lazaric (2008)] learns a set of tasks with different reward functions using the batch method *Fitted Q-iteration* [Ernst et al. (2005)]. By leveraging a multi-task feature learning algorithm [Argyrious et al. (2007)], the problem can be formulated as a joint optimization problem to find the best features and learning parameters across observed data in all tasks. Experiments demonstrate that this method can improve the total reward and can help the agent to ignore irrelevant features (i.e., features which do not provide useful information). Furthermore, since it may be possible to learn a superior representation, asymptotic performance may be improved as well, relative to learning tasks in isolation.

The work in this section, as summarized in the second section of Table 3.1, explicitly assumes that all MDPs an agent experiences are drawn from the same distribution. Different tasks in a single distribution could, in principal, have different state variables and actions, and future work should investigate when allowing such flexibility would be beneficial.

Table 3.5 This table reproduces the third group of methods from Table 3.1

Citation	Allowed Task Differences	Source Task Selection	Task Mappings	Transferred Knowledge	Allowed Learners	TL Metrics
Different state variables and actions – no explicit task mappings: Section 3.4						
[Konidaris and Barto (2006)]	p	h	N/A	R	TD	j, tr
[Konidaris and Barto (2007)]	p	h	N/A	π_p	TD	j, tr
[Banerjee and Stone (2007)]	a, v	h	N/A	fea	TD	ap, j, tr
[Guestrin et al. (2003)]	#	h	N/A	Q	LP	j
[Croonenborghs et al. (2007)]	#	h	N/A	π_p	RRL	ap, j, tr
[Ramon et al. (2007)]	#	h	N/A	Q	RRL	ap, j, tt†, tr
[Sharma et al. (2007)]	#	h	N/A	Q	TD, CBR	j, tr

3.4 Transferring Task-Invariant Knowledge between Tasks with Differing State Variables and Actions

This section, unlike the previous two, discusses methods that allow the source task and target task to have different state variables and actions (see Figure 3.4 and the methods in Table 3.5). These methods formulate the problem so that no explicit mapping between the tasks is needed. Instead the agent reasons over abstractions of the MDP that are invariant when the actions or state variables change.

For example, [Konidaris and Barto (2006)] have separated the standard RL problem into *agent-space* and *problem-space* representations. The agent-space is determined by the agent's capabilities, which remain fixed (e.g., physical sensors and actuators), although such a space may be non-Markovian[3]. The problem-space, on the other hand, may change between source and target problems and is assumed to be Markovian. The authors' method learns a shaping reward on-line in agent-space while learning a source task. If a later target task has a similar reward structure and action set, the learned shaping reward will help the agent achieve a jumpstart and higher total reward. For example, suppose that one of the agent's sensors measures the distance between it and a particular important state (such as a beacon located near the goal state). The agent may learn a shaping reward that assigns reward when the state variable describing its distance to the beacon is reduced, even in the absence of an environmental reward. The authors assume that there are no novel actions (i.e., actions which are not in the source task's problem-space) but any new state variables can be handled if they can be mapped from the novel problem-space into the familiar agent-space. Additionally, the authors acknowledge that the transfer must be between *reward-linked* tasks, where "the reward function in each environment consistently allocates rewards to the same types of interactions across environments," but determining if a sequence of tasks meet this criterion is left for future work.

[3] A standard assumption is that a task is Markovian, meaning that the probability distribution over next states is independent of the agent's state and action history. Thus, saving a history would not assist the agent when selecting actions, and it can consider each state in isolation.

In later work [Konidaris and Barto (2007)], the authors assume knowledge of "pre-specified salient events," which make learning options tractable. While it may be possible to learn options without requiring such events to be specified, the paper focuses on how to use such options rather than option learning. Specifically, when the agent achieves one of these subgoals, such as unlocking a door or moving through a doorway, it may learn an option to achieve the event again in the future. As expected, problem-space options speed up learning a single task. More interesting, when the agent trains on a series of tasks, options in both agent-space and problem-space significantly increase the jumpstart and total reward in the target task (time spent learning the source task is discounted). The authors suggest that agent-space options will likely be more portable than problem-space options in cases where the source and target tasks are less similar—indeed, problem-space options will only be portable when source and target tasks are very similar.

In our opinion, agent- and problem-space are ideas that should be further explored as they will likely yield additional benefits. Particularly in the case of physical agents, it is intuitive that the sensors and actuators of the agent will be static, allowing information such as options to be reused. Task-specific items, such as features and actions, may change, but should be faster to learn if the agent has already learned something about its unchanging agent-space.

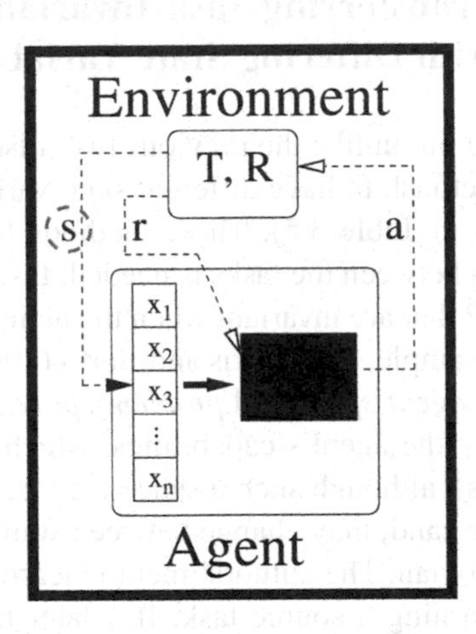

Fig. 3.4 Methods in Section 3.4 are able to transfer between tasks with different state spaces. Although T, R, A, and the state variables may also technically change, the agent's internal representation is formulated so that they remain fixed between source and target tasks. As in previous figures, MDP components with a dashed circle may change between the source task and target task.

If transfer is applied to game trees, changes in actions and state variables may be less problematic. [Banerjee and Stone (2007)] are able to transfer between games by focusing on this more abstract formulation. For instance, in experiments the learner identified the concept of a *fork*, a state where the player could win on the subsequent turn regardless of what move the opponent took next. After training in the source task, analyzing the source task data for such features, and then setting the value for a given feature based on the source task data, such features of the game tree were used in a variety of target tasks. This analysis focuses on the effects of actions on the game tree and thus the actions and state variables describing the source and target game can differ without requiring an inter-task mapping. Source task time is discounted, but jumpstart, total reward, and asymptotic performance are all improved via transfer. Although the experiments in the paper use only temporal difference learning, it is likely that this technique would work well with other types of learners.

[Guestrin et al. (2003)] examine a similar problem in the context of planning in what they term a *relational MDP*. Rather than learning a standard value function, an agent-centered value function for each *class* of agents is calculated in a source task, forcing all agents of a given class type will all have the same value function. However, these class value functions are defined so that they are independent of the number of agents in a task, allowing them to be directly used in a target task which has additional (or fewer) agents. No further learning is done in the target task, but the transferred value functions perform better than a handcoded strategy provided by the authors, despite having additional friendly and adversarial agents. However, the authors note that the technique will not perform well in heterogeneous environments or domains with "strong and constant interactions between many objects."

Relational Reinforcement Learning may also be used for effective transfer. Rather than reasoning about states as input from an agent's sensors, an RRL learner typically reasons about a state in propositional form by constructing first-order rules. The learner can easily abstract over specific object identities as well as the number of objects in the world; transfer between tasks with different number of objects is simplified. For instance, [Croonenborghs et al. (2007)] first learn a source task policy with RRL. The learned policy is used to create examples of state-action pairs, which are then used to build a relational decision tree. This tree predicts, for a given state, which action would be executed by the policy. Lastly, the trees are mined to produce *relational options*. These options are directly used in the target task with the assumption that the tasks are similar enough that no translation of the relational options is necessary. The authors consider three pairs of source/target tasks where relational options learned in the source directly apply to the target task (only the number of objects in the tasks may change), and learning is significantly improved in terms of jumpstart, total reward, and asymptotic performance.

Other work using RRL for transfer [Ramon et al. (2007)] introduces the TGR algorithm, a relational decision tree algorithm. TGR incrementally builds a decision tree in which internal nodes use first-order logic to analyze the current state and where the tree's leaves contain action-values. The algorithm uses four tree-restructuring operators to effectively use available memory and increase sample efficacy. Both target task time and total time are reduced by first training on a simple source task and then on a related target task. Jumpstart, total reward, and asymptotic performance also appear to improve via transfer.

RRL is a particularly attractive formulation in the context of transfer learning. In RRL, agents can typically act in tasks with additional objects without reformulating their knowledge, although additional training may be needed to achieve optimal (or even acceptable) performance levels. When it is possible to frame a domain of interest as an RRL task, transfer between tasks with different numbers of objects or agents will likely be relatively straightforward.

With motivation similar to that of RRL, some learning problems can be framed so that agents choose between high-level actions that function regardless of the number of objects being reasoned about. [Sharma et al. (2007)] combines case-based reasoning with RL in the *CAse-Based Reinforcement Learner* (CARL), a multi-level architecture includes three modules: a planner, a controller, and a learner. The tactical layer uses the learner to choose between high-level actions which are independent of the number of objects in the task. The cases are indexed by: high-level state variables (again independent of the number of objects in the task), the actions available, the Q-values of the actions, and the cumulative contribution of that case on previous timesteps. Similarity between the current situation and past cases is determined by Euclidean distance. Because the state variables and actions are defined so that the number of objects in the task can change, the source and target tasks can have different numbers of objects (in the example domain, the authors use different numbers of player and opponent troops in the source and target tasks). Time spent learning the source task is not counted, but the target task performance is measured in terms of jumpstart, *asymptotic gain* (a metric related to the improvement in average reward over learning), and *overall gain* (a metric based on the total reward accrued).

In summary, methods surveyed in this section all allow transfer between tasks with different state variables and actions, as well as transfer functions, state spaces, and reward functions. By framing the task in an agent-centric space, limiting the domain to game trees, or using a learning method that reasons about variable numbers of objects, knowledge can be transferred between tasks with relative ease because problem representations do not change from the learner's perspective. In general, not all tasks may be formulated so that they conform to the assumptions made by TL methods presented in this section.

3.5 Explicit Mappings to Transfer between Different Actions and State Representations

This section of the chapter focuses on a set of methods which are more flexible than those previously discussed as they allow the state variables and available actions

Table 3.6 This table reproduces the fourth group of methods from Table 3.1

Citation	Allowed Task Differences	Source Task Selection	Task Mappings	Transferred Knowledge	Allowed Learners	TL Metrics
Different state variables and actions – inter-task mappings used: Section 3.5						
[Taylor et al. (2007a)]	a, v	h	sup	Q	TD	tt†
[Taylor et al. (2007b)]	a, v	h	sup	π	PS	tt†
[Taylor et al. (2008b)]	a, v	h	sup	I	MB	ap, tr
[Torrey et al. (2005)] [Torrey et al. (2006)]	a, r, v	h	sup	rule	TD	j, tr
[Torrey et al. (2007)]	a, r, v	h	sup	π_p	TD	j, tr
[Taylor and Stone (2007a)]	a, r, v	h	sup	rule	any/TD	j, tt†, tr
[Taylor and Stone (2007b)]	a, v	h	sup	I	TD, PS	j, tt†, tr

to differ between source and target tasks (see Table 3.6 and Figure 3.5), without the additional restrictions of methods in the previous section. Note that because of changes in state variables and actions, *R*, *S*, and *T*, all technically change as well (they are functions defined over actions and state variables). However, as we elaborate below, some of the methods allow for significant changes in reward functions between the tasks, while most do not.

In [Taylor et al. (2007a)], the authors assume that a mapping between the source and target tasks is provided to the learner. The learner first trains in a source task using a value-function-learning method. Before learning begins in the target task, every action-value for each state in the target task is initialized via learned source task values. This work experimentally demonstrates that value-function transfer can cause significant speedup by transferring between tasks that have different state variables and actions. Additionally, different methods for performing the value-function transfer are examined, different function approximators are successfully used, and multi-step transfer is demonstrated (i.e., transfer from task A to task B to task C). This TL method demonstrates that when faced with a difficult task, it may be faster overall to first train on an artificial source task or tasks and then transfer the knowledge to the target task, rather than training on the target task directly. The authors provide no theoretical guarantees about their method's effectiveness, but hypothesize conditions under which their TL method will and will not perform well, and provide examples of when their method fails to reduce the training time via transfer.

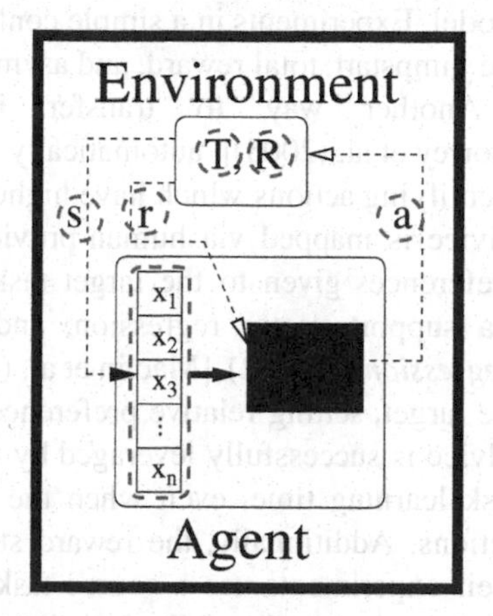

Fig. 3.5 Methods in Section 3.5 focus on transferring between tasks with different state features, action sets, and possible reward functions (which, in turn, causes the state space and transition function to differ as well). As in previous figures, MDP components with a dashed circle may change between the source task and target task.

In subsequent work, [Taylor et al. (2007b)] transfer entire policies between tasks with different state variables and actions, rather than action-value functions. A set of policies is first learned via a genetic algorithm in the source task and then transformed via inter-task mappings. Additionally, partial inter-task mappings are introduced, which may be easier for a human to intuit in many domains. Specifically, those actions and state variables in the target which have "very similar" actions and state variables in the source task are mapped, while novel state variables and actions in the target task are left unmapped. Policies are transformed using one of the inter-task mappings and then used to seed the learning algorithm in the target task. As in the previous work, this TL method can successfully reduce both the target task time and the total time.

Later, [Taylor et al. (2008b)] again consider pairs of tasks where the actions differ, the state variables differ, and inter-task mappings are available to the learner. In this work, the authors allow transfer between model-learning methods by transferring instances, which is similar in spirit to [Lazaric (2008)]. Fitted R-MAX [Jong and Stone (2007)], an instance-based model-learning method capable of learning in continuous state spaces, is used as the base RL method, and source task instances are transferred into the target task to better approximate the target task's model. Experiments in a simple continuous domain show that transfer can improve the jumpstart, total reward, and asymptotic performance in the target task.

Another way to transfer is via learned *advice* or preferences. [Torrey et al. (2005)] automatically extract such advice from a source task by identifying actions which have higher Q-values than other available actions.[4] Such advice is mapped via human-provided inter-task mappings to the target task as preferences given to the target task learner. In this work, Q-values are learned via support vector regression, and then *Preference Knowledge Based Kernel Regression* (KBKR) [Maclin et al. (2005)] adds the advice as soft constraints in the target, setting relative preferences for different actions in different states. The advice is successfully leveraged by the target task learner and decreases the target task learning time, even when the source task has different state variables and actions. Additionally, the reward structure of the tasks may differ substantially: their experiments use a source task whose reward is an unbounded score based on episode length, while the target task's reward is binary, depending on if the agents reached a goal state or not. Source task time is discounted and the target task learning is improved slightly in terms of total reward and asymptotic performance.

Later work [Torrey et al. (2006)] improves upon this method by using *inductive logic programming* (ILP) to identify *skills* that are useful to the agent in a source task. A trace of the agent in the source task is examined and both positive and negative examples are extracted. Positive and negative examples are identified by observing which action was executed, the resulting outcome, the Q-value of the action, and the relative Q-value of other available actions. Skills are extracted using the ILP engine Aleph [Srinivasan (2001)] by using the F_1 score (the harmonic mean of precision and recall). These skills are then mapped by a human into the target task, where they improve learning via KBKR. Source task time is not counted towards the target task time, jumpstart may be improved, and the total reward is improved. The source and target tasks again differ in terms of state variables, actions, and reward structure. The authors also show how human-provided advice may be easily incorporated in addition to advice generated in the source task. Finally, the authors experimentally demonstrate that giving bad advice to the learner is only temporarily harmful and that the learner can "unlearn" bad advice over time, which may be important for minimizing the impact of negative transfer.

[Torrey et al. (2007)] further generalize their technique to transfer *strategies*, which may require composing several skills together, and are defined as a finite-state

[4] While this survey focuses on automatically learned knowledge in a source task, rather than human-provided knowledge, [Torrey et al. (2005)] show that both kinds of knowledge can be effectively leveraged.

machine (FSM). The *structure learning* phase of their algorithm analyzes source task data to find sequences of actions that distinguish between successful and unsuccessful games (e.g., whether or not a goal was reached), and composes the actions into a FSM. The second phase, *ruleset learning*, learns when each action in the strategy should be taken based on state features, and when the FSM should transition to the next state. Experience in the source task is again divided into positive and negative sequences for Aleph. Once the strategies are re-mapped to the target task via a human-provided mapping, they are used to *demonstrate* a strategy to the target task learner. Rather than explore randomly, the target task learner always executes the transferred strategies for the first 100 episodes and thus learns to estimate the Q-values of the actions selected by the transferred strategies. After this demonstration phase, the learner chooses from the MDP's actions, not the high-level strategies, and can learn to improve on the transferred strategies. Experiments demonstrate that strategy transfer significantly improves the jumpstart and total reward in the target task when the source and target tasks have different state variables and actions (source task time is again discounted).

Similar to strategy transfer, [Taylor and Stone (2007a)] learn *rules* with RIPPER [Cohen (1995)] that summarize a learned source task policy. The rules are then transformed via handcoded inter-task mappings so that they could apply to a target task with different state variables and actions. The target task learner may then bootstrap learning by incorporating the rules as an extra action, essentially adding an ever-present option "take the action suggested by the source task policy," resulting in an improved jumpstart and total reward. By using rules as an intermediary between the two tasks, the authors argue that the source and target tasks can use different learning methods, effectively de-coupling the two learners. Additionally, this work demonstrated that *inter-domain* transfer is possible. The two source tasks in this paper were discrete, fully observable, and one was deterministic. The target task, however, had a continuous state space, was partially observable, and had stochastic actions. Because the source tasks required orders of magnitude less time, the total time was roughly equal to the target task time. Similarities with [Torrey et al. (2007)] include a significant improvement in initial performance and no provision to automatically handle scale differences. To our knowledge, there is currently no published method to automatically scale rule constants, which would be necessary if, for instance, the source task's distances were measured in feet, but the target task's distances were measured in meters. The methods differ primarily in how advice is incorporated into the target learner and the choice of utilizing a relatively fast proposition rule learner rather than a first-order rule learner. Although much less source data is needed to learn rules in this setup, the more powerful first order rules need no translation if the objects in question do not change. For example, if a strategy is learned with respect to the distance between teammates, the number of teammates can vary between the source and target without needing to translate such a strategy.

Our past work has used the term "inter-domain transfer" for transfer between qualitatively different domains, such as between a board game and a soccer simulation. However, this term is not well defined, or even agreed upon in the community. For instance, [Swarup and Ray (2006)] use the term "cross-domain transfer" to describe the reuse of a neural network structure between classification tasks with different numbers of boolean inputs and a single output. However, our hope is that researchers will continue improve transfer methods so that they may usefully transfer from very dissimilar tasks, similar to the way that humans may transfer high level ideas between very different domains.

This survey has discussed examples of low- and high-level knowledge transfer. For instance, learning general rules or advice may be seen as relatively high level, whereas transferring specific Q-values or observed instances is quite task-specific. Our intuition is that higher-level knowledge may be more useful when transferring between very dissimilar tasks. For instance, it is unlikely that Q-values learned for a checkers game will transfer to chess, but the concept of a fork may transfer well. This has not been definitely shown, however, nor is there a quantitative way to classify knowledge in terms of low- or high-level. We hope that future work will confirm or disconfirm this hypothesis, as well as generate guidelines as to when different types of transferred knowledge is most appropriate.

Transfer learning problems are typically framed as leveraging knowledge learned on a source task to improve learning on a related, but different, target task. [Taylor and Stone (2007b)] examine the complimentary task of transferring knowledge between agents with different internal *representations* (i.e., the function approximator or learning algorithm) of the *same* task. Allowing for such shifts in representation gives additional flexibility to an agent designer; past experience may be transferred rather than discarded if a new representation is desired. A more important benefit is that changing representations partway through learning can allow agents to achieve better performance in less time. *Complexification* transfers between different function approximator parameterizations, which would typically be used to increase the complexity of a function approximator over time and thus reduce the total required learning time. *Offline Representation Transfer* is used to initialize a second representation using offline training, which can be used to transfer between different function approximators or between different learning methods (e.g., use a learned source task value function to initialize a target task policy search method). Additionally, the authors demonstrate that offline representation transfer can be used for inter-task transfer between tasks with different actions and state variables. Experiments show that both algorithms can reduce the target task training time, improve the jumpstart and total reward, and in some cases reduce the total training time. It is not clear how often this type of method will be useful in practice, but does serve as a proof of concept in that instances can be used to transfer knowledge not only between different tasks, but also between different representations.

All methods in this section use some type of inter-task mapping to allow transfer between MDPs with very different specifications. While these results show that transfer can provide a significant benefit, they presuppose that the mappings are provided to the learner. The following section considers methods that work to autonomously learn such inter-task mappings.

3.6 Learning Task Mappings

The transfer algorithms considered thus far have assumed that a hand-coded mapping between tasks was provided, or that no mapping was needed. In this section we consider methods (the final group in Table 3.1) which learn a mapping between tasks so that source task knowledge may be exploited in a novel target task with different state variables and actions (see Figure 3.6). The space of possible algorithms to learn inter-task mappings is relatively unexplored, compared to methods which can successfully leverage such mappings.

One current challenge of TL research is to reduce the amount of information provided to the learner about the relationship between the source and target tasks. If a human is directing the learner through a series of tasks, the similarities (or analogies) between the tasks will likely be provided by the human's intuition. If transfer is to succeed in an autonomous setting, however, the learner must first determine how (and whether) two tasks are related, and only then may the agent leverage its past knowledge to learn in a target task. Learning task relationships is critical if agents are to transfer without human input, either because the human is outside the loop, or because the human is *unable* to provide similarities between tasks. Methods in this section differ primarily in what information must be provided. At one end of the spectrum, [Kuhlmann and Stone (2007)]

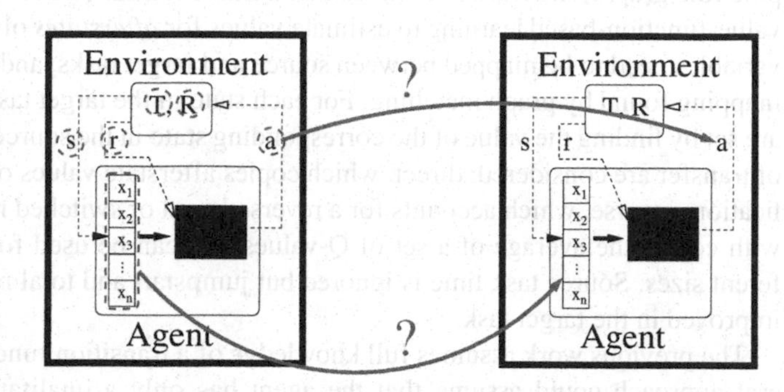

Fig. 3.6 Section 3.6 presents methods to learn the relationship between tasks with different state variables and actions. As in previous figures, MDP components with a dashed circle may change between the source task and target task.

Table 3.7 This table reproduces the group of inter-task learning methods from Table 3.1

Citation	Allowed Task Differences	Source Task Selection	Task Mappings	Transferred Knowledge	Allowed Learners	TL Metrics
Learning inter-task mappings: Section 3.6						
[Kuhlmann and Stone (2007)]	a, v	h	T	Q	TD	j, tr
[Liu and Stone (2006)]	a, v	h	T	N/A	all	N/A
[Soni and Singh (2006)]	a, v	h	M_a, sv_g, exp	N/A	all	ap, j, tr
[Talvitie and Singh (2007)]	a, v	h	M_a, sv_g, exp	N/A	all	j
[Taylor et al. (2007b)]*	a, v	h	sv_g, exp	N/A	all	$tt^\dagger$
[Taylor et al. (2008d)]	a, v	h	exp	N/A	all	j, tr

assume that a complete description of R, S, and T are given, while at the other, [Taylor et al. (2008d)] learn the mapping exclusively from experience gathered via environmental interactions.

Given a complete description of a game (i.e., the full model of the MDP), [Kuhlmann and Stone (2007)] analyze the game to produce a *rule graph*, an abstract representation of a deterministic, full information game. A learner first trains on a series of source task games, storing the rule graphs and learned value functions. When a novel target task is presented to the learner, it first constructs the target task's rule graph and then attempts to find a source task that has an isomorphic rule graph. The learner assumes that a transition function is provided and uses value-function-based learning to estimate values for *afterstates* of games. Only state variables need to be mapped between source and target tasks, and this is exactly the mapping found by graph matching. For each state in the target task, initial Q-values are set by finding the value of the corresponding state in the source task. Three types of transfer are considered: direct, which copies afterstate values over without modification; inverse, which accounts for a reversed goal or switched roles; and average, with copies the average of a set of Q-values and can be used for boards with different sizes. Source task time is ignored but jumpstart and total reward can both be improved in the target task.

The previous work assumes full knowledge of a transition function. A more general approach could assume that the agent has only a qualitative understanding of the transition function. For instance, *qualitative dynamic Bayes networks* (QDBNs) [Liu and Stone (2006)], summarize the effects of actions on state variables but are not precise (for instance, they could not be used as a generative model for planning). If QDBNs are provided to an agent, a graph mapping technique can automatically find a mapping between actions and state variables in two tasks with relatively little computational cost. The authors show that mappings can be learned autonomously, effectively enabling value function transfer between tasks with different state variables and actions. However, it remains an open question as to whether or not QDBNs are learnable from experience, rather than being hand-coded.

The next three methods assume knowledge about how state variables are used to describe objects in a multi-player task. For instance, an agent may know that a pair of state variables describe "distance to teammate" and "distance from teammate to marker," but the agent is not told *which* teammate the state variables describe. First, [Soni and Singh (2006)] supply an agent with a series of possible state transformations and an inter-task action mapping. There is one such transformation, X, for every possible mapping of target task variables to source task variables. After learning the source task, the agent's goal is to learn the correct transformation: in each target task state s, the agent can randomly explore the target task actions, or it may choose to take the action $\pi_{source}(X(s))$. This method has a similar motivation to that of [Fernandez and Veloso (2006)], but here the authors are learning to select between possible mappings rather than possible previous policies. Over time the agent uses Q-learning to select the best state variable mapping as well as learn the action-values for the target task. The jumpstart, total reward, and asymptotic performance are all slightly improved when using this method, but its efficacy will be heavily dependent on the number of possible mappings between any source and target task.

Second, *AtEase* [Talvitie and Singh (2007)] also generates a number of possible state variable mappings. The action mapping is again assumed and the target task learner treats each of the possible mappings as an arm on a multi-armed bandit [Bellman (1956)]. The authors prove that their algorithm learns in time proportional to the number of possible mappings rather than the size of the problem: "in time polynomial in T, [the algorithm] accomplishes an actual return close to the asymptotic return of the best expert that has mixing time at most T." This approach focuses on efficiently selecting between the proposed state variable mappings and does not allow learning in the target task.

Third, these assumptions are relaxed slightly by [Taylor et al. (2007b)], who show that it is possible to learn both the action and state variable mapping simultaneously by leveraging a classification technique, although it again relies on the pre-specified state variable groupings (i.e., knowing that "distance to teammate" refers to a teammate, but not which teammate). Action and state variable classifiers are trained using recorded source task data. For instance, the source task agent records s_{source}, a_{source}, s'_{source} tuples as it interacts with the environment. An action classifier is trained so that $C(s_{source,object}, s'_{source,object}) = a_{source}$ for each object present in the source task. Later, the target task agent again records s_{target}, a_{target}, s'_{target} tuples. Then the action classifier can again be used for to classify tuples for every target task object: $C(s_{target,object}, s'_{target,object}) = a_{source}$, where such a classification would indicate a mapping between a_{target} and a_{source}. Relatively little data is needed for accurate classification; the number of samples needed to learn in the target task far outweighs the number of samples used by the mapping-leaning step. While the resulting mappings are not always optimal for transfer, they do serve to effectively reduce target task training time as well as the total training time.

The MASTER algorithm [Taylor et al. (2008d)] was designed to further relax the knowledge requirements of [Taylor et al. (2007b)]: no state variable groupings are required. The key idea of MASTER is to save experienced source task instances, build an approximate transition model from a small set of experienced target task instances, and then test possible mappings offline by measuring the prediction error of the target-task models on source task data. This approach is sample efficient at the expense of high computational complexity, particularly as the number of state variables and actions increase. The method uses an exhaustive search to find the inter-task mappings that minimize the prediction error, but more sophisticated (e.g., heuristic) search methods could be incorporated. Experiments show that the learned inter-task mappings can successfully improve jumpstart and total reward. A set of experiments also shows how the algorithm can assist with source task selection by selecting the source task which is best able to minimize the offline prediction error. The primary contribution of MASTER is to demonstrate that autonomous transfer is possible, as the algorithm can learn inter-task mappings autonomously, which may then be used by any of the TL methods discussed in the previous section of this chapter (Section 3.5).

In summary, this last section of the chapter has discussed several methods able to learn inter-task mappings with different amounts of data. Although all make some assumptions about the amount of knowledge provided to the learner or the similarity between source and target tasks, these approaches represent an important step towards achieving fully autonomous transfer.

The methods in the section have been loosely ordered in terms of increasing autonomy. By learning inter-task mappings, these algorithms try to enable a TL agent to use past knowledge on a novel task without human intervention, even if the state variables or actions change. However, the question remains whether fully autonomous transfer would ever be useful in practice. Specifically, if there are no restrictions on the type of target task that could be encountered, why would one expect that past knowledge (a type of bias) would be useful when learning an encountered task, or even on the majority of tasks that could be encountered? This question is directly tied to the ability of TL algorithms to recognize when tasks are similar and when negative transfer may occur, both of which are discussed in more detail in the following section.

Chapter 4
Empirical Domains

In this chapter we introduce the testbed domains used in the remainder of this monograph, where we informally define a domain as a setting for one or more tasks (i.e., MDPs). The primary purpose of this chapter is to provide sufficient background to allow a reader to fully understand the transfer learning experiments in subsequent chapters. In addition to describing each domain, we explain how each can be learned[1] with one or more of the RL methods discussed in the previous chapter. In every domain we will emphasize how some tasks are faster to master than others. In general, experiments in this monograph transfer from relatively simple, quick to learn, source task to a more complex target task. In the target task time scenario, an effective TL algorithm may reduce the target task learning time regardless of task ordering, but only by ordering tasks in order of increasing difficulty will the total training time be reduced.

The canonical Mountain Car task and an extension named 3D Mountain Car are discussed in Section 4.1. Section 4.2 presents Server Job Scheduling, a discrete task with many actions. Keepaway is the most complex domain used in this monograph and is presented in Section 4.3. We then introduce two tasks that have been explicitly constructed to be similar to Keepaway: Ringworld (Section 4.4) and Knight Joust (Section 4.5). Finally, in Section 4.6, we summarize and contrast the domains in terms of their properties and the different challenges learners face when acting in a task drawn from each domain.

4.1 Generalized Mountain Car

This section introduces a generalized version of the standard RL benchmark *Mountain Car* task [Moore (1991), Singh and Sutton (1996)]. Mountain Car is among the simplest continuous domains that can benefit from RL, and it is generalizable to be appropriate for TL experiments.

---

[1] Recall that because this monograph is about transfer learning, not about base learning algorithms, we assume that some type of learning is possible in the experimental domains.

M.E. Taylor: Transfer in Reinforcement Learning Domains, SCI 216, pp. 61–90.
springerlink.com

In the standard 2D Mountain Car task, the agent must generalize across continuous state variables in order to drive an underpowered car up a mountain to a goal state. We also introduce a 3D Mountain Car task [Taylor et al. (2008d)] as extension of the 2D task, retaining much of the structure of the 2D problem. In both tasks the transition and reward functions are initially unknown. The agent begins at rest at the bottom of the hill and receives a reward of -1 at each time step. After reaching a goal state or taking 500 actions, whichever comes first, the agent is reset to the start state.

4.1.1 Two Dimensional Mountain Car

In the two dimensional Mountain Car task (mentioned previously in Section 3.1), the car travels along the curve $\sin(3x)$ between $-1.2 \leq x \leq 0.6$. Two continuous variables fully describe the agent's state: The horizontal position, x, and velocity, $\dot{x}$, which has a range of $[-0.07, 0.07]$. The agent may select one of three actions at each timestep; the actions {`Left`, `Neutral`, and `Right`} change the velocity by -0.001, 0, and 0.001 respectively. The state variables and actions are listed in Table 4.1. Additionally, $-0.025(\cos(3x))$ is added to $\dot{x}$ at each timestep to account for the force of gravity on the car.[2] The start state is at the bottom of the hill ($x = -\frac{\pi}{6}, \dot{x} = 0$), and the goal states are those where $x \geq 0.5$ (see Figure 4.1). We use the publicly available version of this code for our experiments.[3] Figure 4.2 shows an example learned policy.

The transfer experiments in this monograph use four variants of the 2D Mountain Car task. The No Goal 2D task, Low Power 2D task, and High Power 2D task are novel tasks created to investigate transfer efficacy.

- The *Standard 2D task* is described in the previous paragraph and is a standard RL benchmark task.
- The *No Goal 2D task* is the same as the standard task, except that goal state has been removed. Every episode lasts 500 timesteps.
- The third variant, *Low Power 2D task*, changes the car so that actions modify the velocity by ± 0.0007, making the task more difficult than the Standard 2D task.
- The fourth variant, *Hand Brake 2D task*, adds a fourth action to the Low Power 2D task which sets the car's velocity to zero. This is particularly problematic for the agent because it must learn that this action is never useful (and, when executed, makes it more difficult for the car to reach the goal because all kinetic energy is lost).
- The fifth variant is the *High Power 2D task*, where actions change the car's velocity by ± 0.0015: the car has 50% more acceleration than the benchmark car.

[2] The acceleration due to gravity of the car is determined by the slope of the surface at the car's current location, found by taking the derivative of the $\sin(3x)$ curve, which gives a constant factor multiplied by $\cos(3x)$.

[3] Available as a domain for use in the RL-Glue framework at: `http://rlai.cs.ualberta.ca/RLR/MountainCarBestSeller.html`

Table 4.1 This table summarizes the state space and the action space of the 2D Mountain Car task

	State Variables	Actions	
2D Mountain Car	x position	`Left`	accelerate left
	$\dot{x}$ velocity	`Neutral`	no acceleration
		`Right`	accelerate right

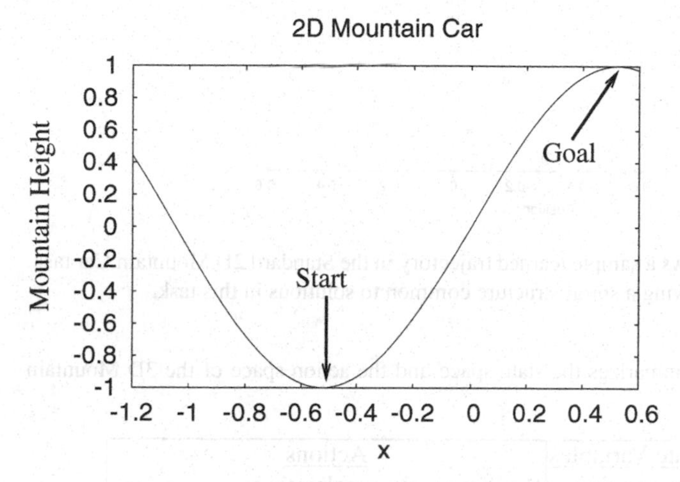

Fig. 4.1 In the 2D Mountain Car task, the agent must travel along a curve (mountain)

In Section 7.1.3, the No Goal 2D task will be used to show how the effectiveness of transfer changes when using a different reward function, R. The other three variants will be used to show how transfer efficacy changes when a source task's transition function changes (relative to the Standard 2D task).

4.1.2 Three Dimensional Mountain Car

To create a three dimensional task, we extend the mountain's curve into a 3D surface (see Figure 4.3) defined by $\sin(3x) + sin(3y)$. The state is composed of four continuous state variables: $x, \dot{x}, y, \dot{y}$. The positions and velocities have ranges of $[-1.2, 0.6]$ and $[-0.07, 0.07]$, respectively. The agent selects from five actions at each timestep: {`Neutral`, `West`, `East`, `South`, and `North`} (summarized in Table 4.2). `West` and `East` modify $\dot{x}$ by -0.001 and +0.001 respectively, while `South` and `North` modify $\dot{y}$ by -0.001 and +0.001 respectively.[4] The force of gravity adds $-0.025(\cos(3x))$ and $-0.025(\cos(3y))$ to $\dot{x}$ and $\dot{y}$, respectively, at each time step,

[4] Although we call the agent's vehicle a "car," it does not turn but simply accelerates in the four cardinal directions. In Moore's original paper, the 2D task was called the "puck-on-hill."

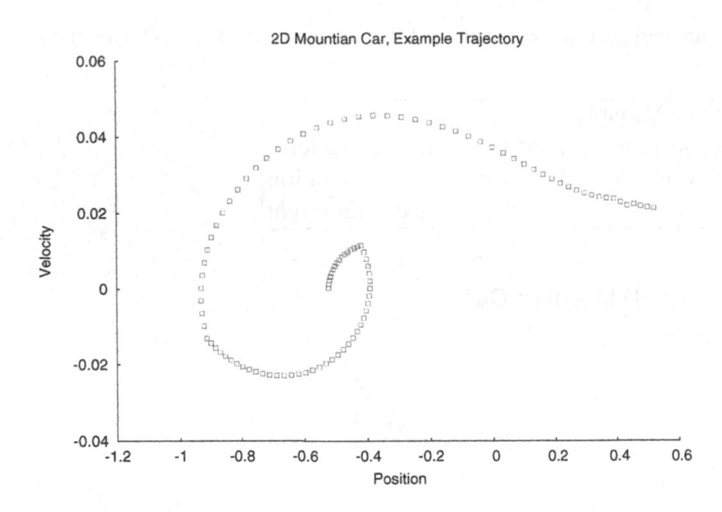

Fig. 4.2 This figure shows a sample learned trajectory in the Standard 2D Mountain Car task. x and $\dot{x}$ are graphed showing a spiral structure common to solutions in this task.

Table 4.2 This table summarizes the state space and the action space of the 3D Mountain Car task

	State Variables	Actions	
3D Mountain Car	x x-position	`Neutral`	no acceleration
	y y-position	`West`	accelerate in the -x direction
	$\dot{x}$ x-velocity	`East`	accelerate in the x direction
	$\dot{y}$ y-velocity	`South`	accelerate in the -y direction
		`North`	accelerate in the y direction

where the force of gravity is again proportional to the slope of the surface at the agent's position. The goal state region is defined by $x \geq 0.5$ and $y \geq 0.5$.

This task is more difficult than the 2D task because of the increased number of state variables and additional actions (see Figure 4.4 for an example trajectory). Furthermore, since the agent can affect its acceleration in only one of the two spacial dimensions at any given time, this problem cannot readily be "factored" into the simpler 2D task. While data gathered from a 2D task should be able to help an agent learn a 3D Mountain Car task, we do expect that some amount of learning would be required even after transferring information from a source task.

In addition to the *Standard 3D task* described above, we also define a *Low Power 3D task*, and a *Hand Brake 3D task*. In the low power task, each action modifies the velocities by ± 0.0007, which makes discovering the goal state more difficult. The hard brake variant adds a 6th action which has the effect of setting the car's velocity to zero and makes discovering the goal more difficult.

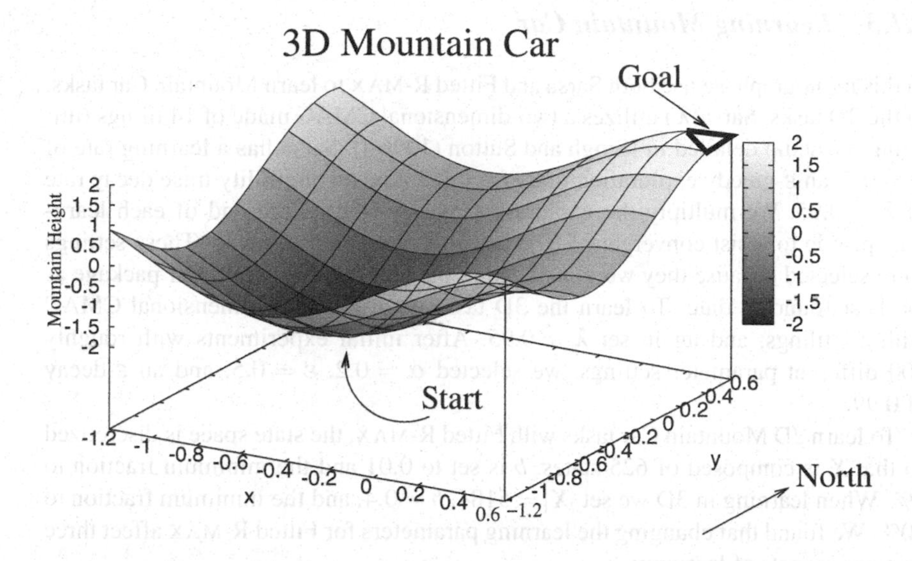

Fig. 4.3 In 3D Mountain Car the 2D curve becomes a 3D surface. The agent starts at the bottom of the hill with no kinetic energy and attempts to reach the goal area in the Northeast corner.

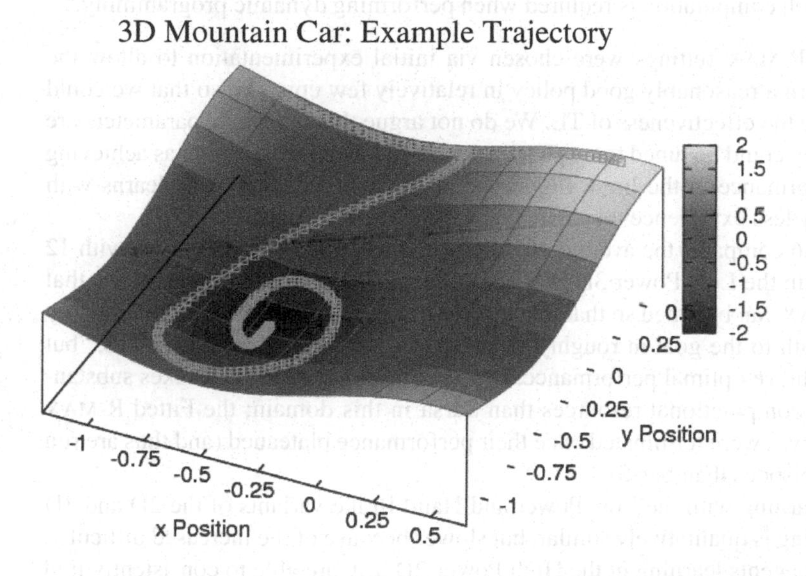

Fig. 4.4 This figure shows a sample learned trajectory in the 3D Mountain Car task. x and y are graphed and $\dot{x}$ and $\dot{y}$ can be inferred by the spacing between the samples in the trajectory, plotted once per timestep.

4.1.3 Learning Mountain Car

In this monograph we use both Sarsa and Fitted R-MAX to learn Mountain Car tasks. In the 2D tasks, Sarsa(λ) utilizes a two-dimensional CMAC made of 14 tilings (utilizing the setup detailed in [Singh and Sutton (1996)]). Sarsa has a learning rate of $\alpha = 0.5$, an ε-greedy exploration rate of $\varepsilon = 0.1$, and an eligibility trace decay rate of $\lambda = 0.95$. We multiply the exploration rate by 0.99 at the end of each learning episode to assist convergence. The learning rate is not decayed. These settings were selected because they were included in the released Mountain Car package as the best found to date. To learn the 3D task, we use a four-dimensional CMAC with 14 tilings, and again set $\lambda = 0.95$. After initial experiments with roughly 100 different parameter settings, we selected $\alpha = 0.2$, $\varepsilon = 0.5$, and an ε-decay of 0.99.

To learn 2D Mountain Car tasks with Fitted R-MAX, the state space is discretized so that X is composed of 625 states. b is set to 0.01 and the minimum fraction to 1%. When learning in 3D we set $|X| = 5184$, $b = 0.4$, and the minimum fraction to 10%. We found that changing the learning parameters for Fitted R-MAX affect three primary aspects of learning:

- How accurately the optimal policy can be approximated.
- How many samples are needed to accurately approximate the best policy, given the representation.
- How much computation is required when performing dynamic programming.

The Fitted R-MAX settings were chosen via initial experimentation to allow the agent to learn a reasonably good policy in relatively few episodes so that we could demonstrate the effectiveness of TL. We do not argue that the above parameters are optimal; they could be tuned to emphasize any of the above goals, such as achieving higher performance in the limit. Figure 4.5 shows that Fitted R-MAX learns with significantly less experience than Sarsa on the Standard 2D task.

Figure 4.6 compares the average performance of 12 Fitted R-MAX trials with 12 Sarsa trials in the Low Power 3D Mountain Car task. This result demonstrates that Fitted R-MAX can be tuned so that it learns with significantly less data, consistently finding a path to the goal in roughly 50 episodes instead of 10,000 episodes,[5] but does not achieve optimal performance. Learning with Fitted R-MAX takes substantially more computational resources than Sarsa in this domain; the Fitted R-MAX learning curves were terminated once their performance plateaued (and thus are run for fewer episodes than Sarsa).

When learning with the Low Power and Hand Brake variants of the 2D and 3D tasks, learning is qualitatively similar, but slower because of the increased difficulty. Conversely, agents learning in the High Power 2D task are able to consistently find the goal faster than in the Normal or Low Power tasks. Finally, because the NoGoal

[5] We hypothesize this is due in part to Fitted R-MAX's efficient exploration scheme, which may allow the agent to discover the goal much faster than Sarsa's random exploration.

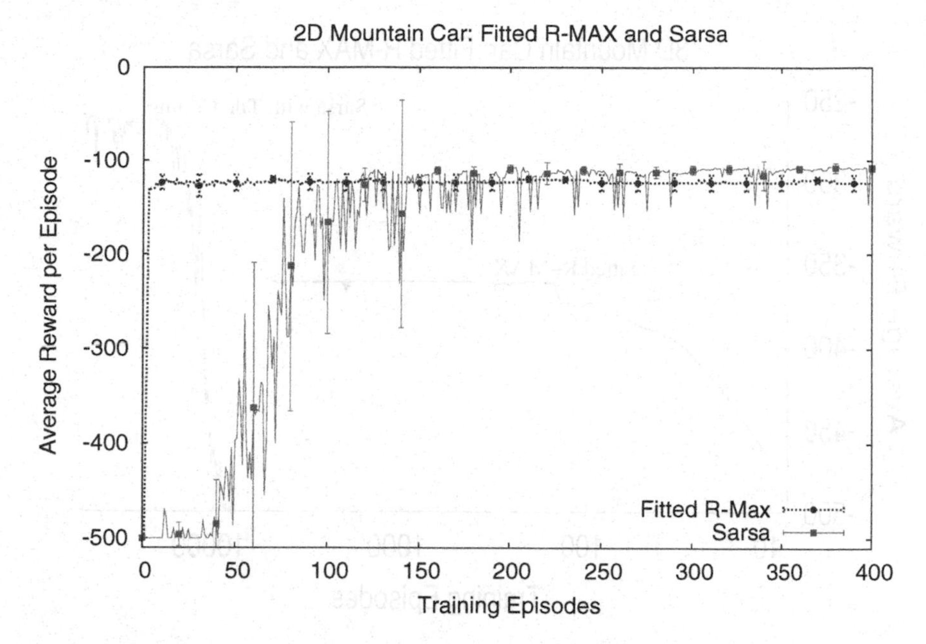

Fig. 4.5 Average learning curves for Fitted R-MAX and Sarsa show that Fitted R-MAX initially learns to find the goal consistently at the expense of a slightly lower asymptotic performance. Error bars show the standard deviation of each experiment (composed of 12 independent learning trials).

task has a uniform reward function and every episode lasts for 500 episodes, agents learning in this task cannot affect their average reward over time.

Transfer experiments later in this monograph will use 2D Mountain Car tasks as source tasks, and 3D Mountain Car tasks as target tasks. The reader may have noticed that the Standard 2D task (Figure 4.5) is learned much faster than the Standard 3D task (Figure 4.6) for both Fitted R-MAX and Sarsa. It is precisely this difference in task difficulty that TL methods can leverage to reduce the total training time, first learning a source task and then a target task, rather than directly learning a target task.

4.2 Server Job Scheduling

While Mountain Car tasks require an agent to learn in a continuous state space, *Server Job Scheduling* [Whiteson and Stone (2006)] (SJS) has a discrete state space but many more available actions. SJS is an *autonomic computing* [Kephart and Chess (2003)] control task in which a server, such as a website's application server or database, must determine in what order to process jobs waiting in its queue. The agent's goal is to maximize the aggregate utility of all the jobs it processes. A *utility function* for each job type maps the job's completion time to the

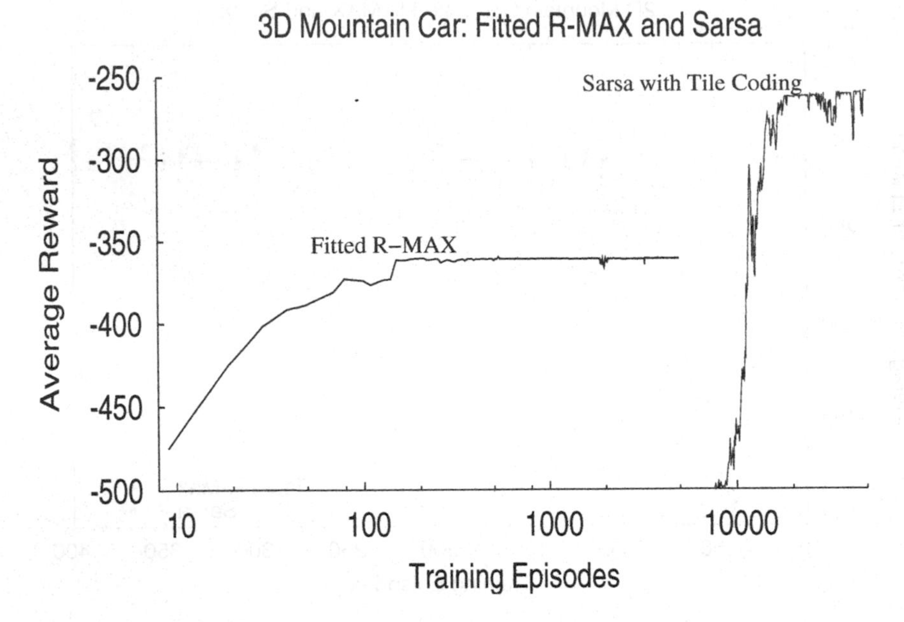

Fig. 4.6 Average learning curves for Fitted R-MAX and Sarsa show the significant advantage, in terms of sample complexity, of model-based RL on the Low Power 3D Mountain Car task. (Note that the x-axis uses a log scale.)

utility derived by the user [Walsh et al. (2004)]. This type of scheduling problem becomes challenging when these utility functions are non-linear and/or the server must process multiple types of jobs. Since selecting a particular job for processing necessarily delays the completion of all other jobs in the queue, the scheduler must weigh difficult trade-offs to maximize aggregate utility.

Each experiment in our simulator begins with 100 jobs pre-loaded into the server's queue and ends when the queue becomes empty. During each timestep, the server removes and processes one job from its queue. For each of the first 100 timesteps, a new job of a randomly selected type is added to the beginning of the queue after the agent processes a job, forcing the agent to make decisions about which job to process as new jobs are arriving. If a job is located at the end of the queue, the server must process the job, limiting each job's life to 200 timesteps. Each episode lasts 200 timesteps. The scheduling agent receives an immediate reward for each job that completes, as determined by that job's age and utility function.

Utility functions for the four job types used in our experiments, which are not provided to the scheduling agent, are shown in Figure 4.7. We consider two tasks in the SJS domain. The *2-job-type SJS task* requires the scheduler to select jobs of types #1 and #2, and the *4-job-type SJS task* uses job types #1–#4. Learning to

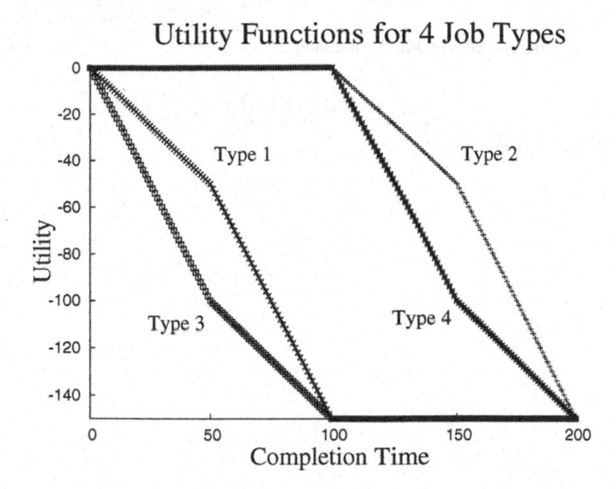

Fig. 4.7 The four utility functions used in our SJS experiments

Table 4.3 This table summarizes the state space and the action space of the 2-job-type Server Job Scheduling task. The 4-job-type task has twice the number of state variables and actions.

2-job-type SJS Task

	State Variables		Actions
$C_{1,1}$	Count of type 1 jobs aged 1-50	$P_{1,1}$	Process oldest type 1 job aged 1-50
$C_{1,2}$	Count of type 1 jobs aged 51-100	$P_{1,2}$	Process oldest type 1 job aged 51-100
$C_{1,3}$	Count of type 1 jobs aged 101-150	$P_{1,3}$	Process oldest type 1 job aged 101-150
$C_{1,4}$	Count of type 1 jobs aged 151-200	$P_{1,4}$	Process oldest type 1 job aged 151-200
$C_{2,1}$	Count of type 2 jobs aged 1-50	$P_{2,1}$	Process oldest type 2 job aged 1-50
$C_{2,2}$	Count of type 2 jobs aged 51-100	$P_{2,2}$	Process oldest type 2 job aged 51-100
$C_{2,3}$	Count of type 2 jobs aged 101-150	$P_{2,3}$	Process oldest type 2 job aged 101-150
$C_{2,4}$	Count of type 2 jobs aged 151-200	$P_{2,4}$	Process oldest type 2 job aged 151-200

schedule with only two task types is significantly easier than learning to schedule all four types.

State and action spaces are discretized: the range of possible job ages is divided into four equal sections. At each timestep, the scheduler knows how many jobs of each type in the queue fall into each range, resulting in 8 state variables in the 2-job-type task and 16 in the 4-job-type SJS task. The action space is similarly discretized to 8 or 16 distinct actions: rather than selecting a particular job, the scheduler specifies what type of job to process and which of the four age ranges that job should lie in. See Table 4.3 for a list of the state variables and actions in the 2-job-type SJS task.

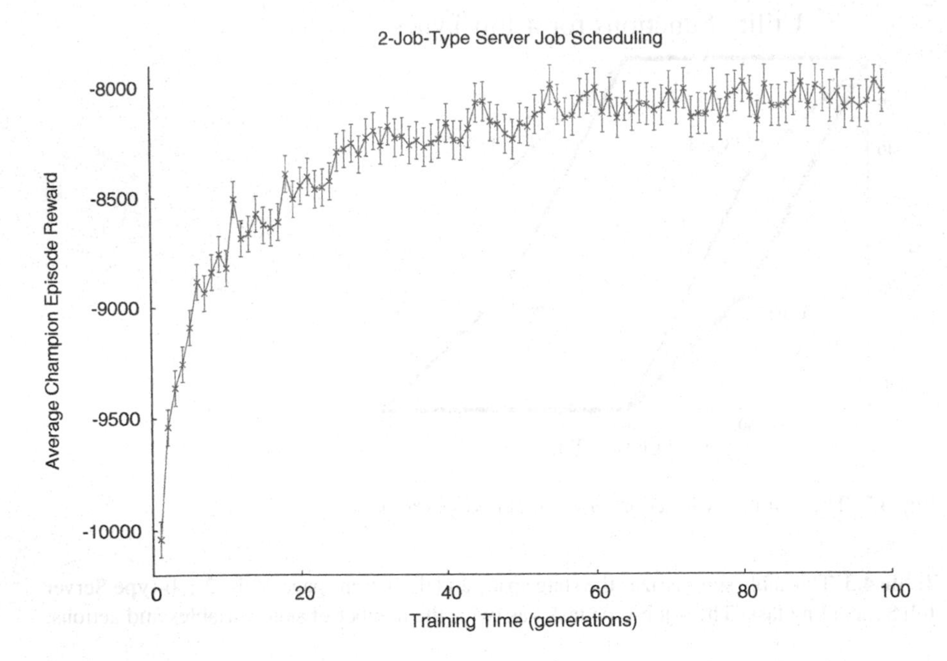

Fig. 4.8 This graph shows that NEAT can successfully learn the 2-job-type SJS task. The x-axis shows the training time in generations. In each generation, the agent evaluates 50 policies for 5 episodes, resulting in 250 episodes per generation. Error bars show the standard deviation of the champion performance per generation, averaged over ten learning curves.

4.2.1 Learning Server Job Scheduling

To learn the 2- and 4-job-type SJS tasks we utilize NEAT, as initial experiments showed it significantly outperformed Sarsa with CMAC function approximation, perhaps due to the large conjunctive state space. Each NEAT network has 8 inputs and outputs in the 2-job-type task and 16 in the 4-job-type task. We use a population size of 50 and evaluate each policy by averaging the reward from 5 episodes. After training is finished, the champion policies are evaluated for an additional 95 episodes to reduce noise when graphing performance. This learning setup is similar to past research in this domain, and all other NEAT parameters are the same as those reported in [Whiteson and Stone (2006)]. Figures 4.8 and 4.9 show the performance of NEAT when learning 2- and 4-job-type SJS tasks. The learning curves show that NEAT can successfully learn in this domain. [Whiteson and Stone (2006)] show that 4-job-type task schedules learned via NEAT outperform three other scheduling methods and compare favorably to a near-optimal schedule (found by solving a large linear program on each timestep).

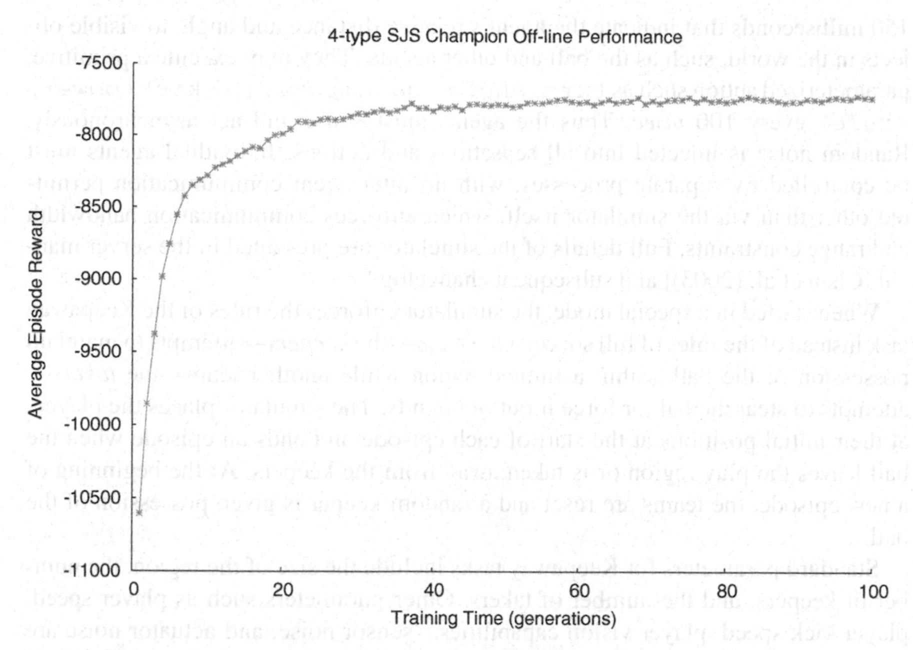

Fig. 4.9 This graph shows that NEAT can successfully learn the 4-job-type SJS task. The x-axis shows the training time in generations. In each generation, the agent evaluates 50 policies for 5 episodes, resulting in 250 episodes per generation.

4.3 Robot Soccer Keepaway

In this section we discuss *Keepaway*, a domain with a continuous state space and significant amounts of noise in the agent's actions and sensors. Keepaway is situated in RoboCup simulated soccer, which has been the basis of multiple international competitions and research challenges. The multiagent domain incorporates noisy sensors and actuators, as well as enforcing a hidden state so that agents only have a partial world view at any given time. While previous work has attempted to use machine learning to learn the full simulated soccer problem [Andre and Teller (1999), Riedmiller et al. (2001)], the complexity and size of the problem have so far proven intractable. However, many RoboCup subproblems have been isolated and solved using machine learning techniques, including the task of playing Keepaway. By focusing on the smaller task of Keepaway, we are able to use RL to learn an action-value function for this complex task and hold the required computational resources to manageable levels.

Since late 2002, the Keepaway task has been part of the official release of the open source RoboCup Soccer Server used at RoboCup (starting with version 9.1.0). In our experiments, unless noted otherwise, we use version 9.4.5 of the RoboCup Soccer Server. Unless direct comparisons with previous work is needed, researchers are encouraged to use version 11.1.0 or higher, which fixed a significant memory leak. Agents in the simulator [Noda et al. (1998)] receive visual perceptions every

150 milliseconds that indicate the agent's relative distance and angle to visible objects in the world, such as the ball and other agents. They may execute a primitive, parameterized action such as `turn(`*`angle`*`)`, `dash(`*`power`*`)`, or `kick(`*`power, angle`*`)` every 100 *msec*. Thus the agents must sense and act asynchronously. Random noise is injected into all sensations and actions. Individual agents must be controlled by separate processes, with no inter-agent communication permitted other than via the simulator itself, which enforces communication bandwidth and range constraints. Full details of the simulator are presented in the server manual [Chen et al. (2003)] and subsequent changelog.[6]

When started in a special mode, the simulator enforces the rules of the Keepaway task instead of the rules of full soccer. One team—the *keepers*—attempts to maintain possession of the ball within a limited region while another team—the *takers*—attempts to steal the ball or force it out of bounds. The simulator places the players at their initial positions at the start of each episode and ends an episode when the ball leaves the play region or is taken away from the keepers. At the beginning of a new episode, the teams are reset and a random keeper is given possession of the ball.

Standard parameters for Keepaway tasks include the size of the region, the number of keepers, and the number of takers. Other parameters such as player speed, player kick speed, player vision capabilities,[7] sensor noise, and actuator noise are also adjustable. Keepaway tasks described in this monograph use standard settings, with the exception of a task described in Section 4.3.3 that uses a different pass actuator. Figure 4.10 shows a diagram of *3 vs. 2 Keepaway* (3 keepers and 2 takers).[8]

When Keepaway was introduced as a testbed [Stone et al. (2005)], a standard task was defined. Most experiments are run on a code base derived from version 0.5 of the benchmark Keepaway implementation.[9] Exceptions include some of the Keepaway illustrative experiments in this chapter, and the experiments in Section 7.2, which use version 0.6 (the newest version of the players that includes some bug fixes). The text will note if version 0.5 is not used, and all comparisons made in this monograph use the players with the same code base.

While some recent work [Iscen and Erogul (2008)] allows takers to learn, our setup is similar to past research in Keepaway [Stone et al. (2005)]; we concentrate exclusively on keeper learning and assume that takers do not learn. Keepers are initially placed near three corners of the square field and a ball is placed near one of the keepers. The two takers are placed in the fourth corner. When the episode

[6] Changelogs are text files describing all changes made to the Soccer Server and are distributed with the server source code.

[7] In our experiments we set the players to have a 360° field of view. Although agents also learn with a more realistic 90° field of view, allowing the agents to see 360° speeds up the rate of learning, enabling more experiments. Additionally, 360° vision also increases the learned performance when compared to learning with the limited 90° vision.

[8] Flash files illustrating the task are available at:
`http://www.cs.utexas.edu/~AustinVilla/sim/Keepaway/`.

[9] Released players are available at
`http://www.cs.utexas.edu/~AustinVilla/sim/Keepaway/`.

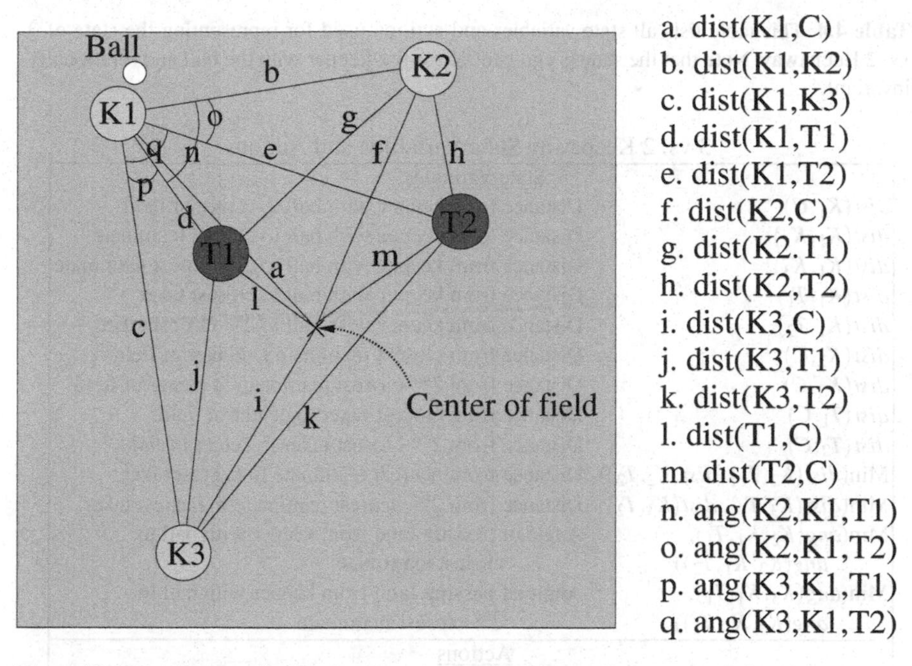

Fig. 4.10 This diagram depicts the distances and angles used to construct the 13 state variables used for learning with 3 keepers and 2 takers. Relevant objects are the 5 players, ordered by distance from the ball, and the center of the field, C. All 13 state variables are enumerated in Table 4.4.

starts, the three keepers attempt to maintain control of the ball by passing among themselves and moving to open positions.

The keeper with the ball has the option to either pass the ball to one of its two teammates or to hold the ball. In this task $A = \{$`hold, pass to closest teammate, pass to second closest teammate`$\}$. S is defined by 13 state variables, enumerated in Table 4.4. The reward to the learning algorithm is the number of time steps the ball remains in play after an action is taken. The keepers learn in a constrained policy space: they have the freedom to decide which action to take only when in possession of the ball. Keepers not in possession of the ball are required to execute the `Receive` macro-action in which the player who can reach the ball the fastest goes to the ball and the remaining players follow a handcoded strategy to try to get open for a pass.

4.3.1 More Complex Keepaway Tasks

As more players are added to the task, Keepaway becomes harder for the keepers because the field becomes more crowded. As more takers are added, there are more opponents to block passing lanes and chase down errant passes. As more keepers are added, the keeper with the ball has more passing options, but the average pass

Table 4.4 This table lists all state variables and actions used for representing the state of 3 vs. 2 Keepaway. Note that the state is ego-centric for the keeper with the ball and rotationally invariant.

3 vs. 2 Keepaway State Variables and Actions

	State Variables
$dist(K_1, C)$	Distance from keeper with ball to center of field
$dist(K_1, K_2)$	Distance from keeper with ball to closest teammate
$dist(K_1, K_3)$	Distance from keeper with ball to 2nd closest teammate
$dist(K_1, T_1)$	Distance from keeper with ball to closest taker
$dist(K_1, T_2)$	Distance from keeper with ball to 2nd closest taker
$dist(K_2, C)$	Distance from closest teammate to center of field
$dist(K_3, C)$	Distance from 2nd closest teammate to center of field
$dist(T_1, C)$	Distance from closest taker to center of field
$dist(T_2, C)$	Distance from 2nd closest taker to center of field
$\text{Min}(dist(K_2, T_1), dist(K_2, T_2))$	Distance from nearest teammate to nearest taker
$\text{Min}(dist(K_3, T_1), dist(K_3, T_2))$	Distance from 2nd nearest teammate to nearest taker
$\text{Min}(ang(K_2, K_1, T_1), ang(K_2, K_1, T_2))$	Angle of passing lane from keeper with ball to closest teammate
$\text{Min}(ang(K_3, K_1, T_1), ang(K_3, K_1, T_2))$	Angle of passing lane from keeper with ball to 2nd closest teammate
	Actions
`Hold`	Maintain possession of the ball (as far as possible from nearest taker)
`Pass`$_1$	Pass to closest teammate
`Pass`$_2$	Pass to second-closest teammate

distance is shorter. This reduced distance forces more passes and often leads to more errors because of the noisy actuators and sensors. Additionally, because each keepers learns independently and a keeper can only select actions when it controls the ball, adding more keepers to the task means that keepers receive less experience per timestep. For these reasons, keepers in *4 vs. 3 Keepaway* require more data to learn a good control policy, compared to 3 vs. 2 Keepaway. The average episode length of the best policy for a constant field size decreases when adding an equal number of keepers and takers, and the time needed to learn a policy with performance equal to a handcoded solution roughly doubles with each additional keeper and taker [Stone et al. (2005)].

In the 4 vs. 3 task, all three takers again start the episode in a single corner. Three keepers start in each of the other three corners and the fourth keeper begins each episode at the center of the field. The reward function is effectively unchanged from 3 vs. 2 Keepaway because the agents still receive a reward of +1 on each timestep, and the transition function is similar because the simulator is unchanged. Now $A = \{$`hold`, `pass to closest teammate`, `pass to second closest teammate`, `pass to third closest teammate`$\}$, and S is made up of 19 state variables due to the added players (see Table 4.5).

Table 4.5 This table lists all state variables and actions used in 4 vs. 3 Keepaway. Novel state variables and actions resulting from the addition of K_4 and T_3 are in **bold.**

4 vs. 3 Keepaway State Variables and Actions

State Variables	
$dist(K_1,C)$	Distance from keeper with ball to center of field
$dist(K_1,K_2)$	Distance from keeper with ball to closest teammate
$dist(K_1,K_3)$	Distance from keeper with ball to 2nd closest teammate
$\mathbf{dist(K_1,K_4)}$	Distance from keeper with ball to 3rd closest teammate
$dist(K_1,T_1)$	Distance from keeper with ball to closest taker
$dist(K_1,T_2)$	Distance from keeper with ball to 2nd closest taker
$\mathbf{dist(K_1,T_3)}$	Distance from keeper with ball to 3rd closest taker
$dist(K_2,C)$	Distance from closest teammate to center of field
$dist(K_3,C)$	Distance from 2nd closest teammate to center of field
$\mathbf{dist(K_4,C)}$	Distance from 3rd closest teammate to center of field
$dist(T_1,C)$	Distance from closest taker to center of field
$dist(T_2,C)$	Distance from 2nd closest taker to center of field
$\mathbf{dist(T_3,C)}$	Distance from 3rd closest taker to center of field
$\mathbf{Min(dist(K_2,T_1), dist(K_2,T_2), dist(K_2,T_3))}$	Distance from nearest teammate to nearest taker
$\mathbf{Min(dist(K_3,T_1), dist(K_3,T_2), dist(K_3,T_3))}$	Distance from 2nd nearest teammate to nearest taker
$\mathbf{Min(dist(K_4,T_1), dist(K_4,T_2), dist(K_4,T_3))}$	Distance from 3rd nearest teammate to nearest taker
$\mathbf{Min(ang(K_2,K_1,T_1), ang(K_2,K_1,T_2), ang(K_2,K_1,T_3))}$	Angle of passing lane from keeper with ball to closest teammate
$\mathbf{Min(ang(K_3,K_1,T_1), ang(K_3,K_1,T_2), ang(K_3,K_1,T_3))}$	Angle of passing lane from keeper with ball to 2nd closest teammate
$\mathbf{Min(ang(K_4,K_1,T_1), ang(K_4,K_1,T_2), ang(K_4,K_1,T_3))}$	Angle of passing lane from keeper with ball to 3rd closest teammate
	Actions
Hold	Maintain possession of the ball (as far as possible from nearest taker)
$\mathtt{Pass}_1$	Pass to closest teammate
$\mathtt{Pass}_2$	Pass to second-closest teammate
$\mathbf{Pass_3}$	Pass to third-closest teammate

It is also important to point out that the addition of an extra taker and keeper in 4 vs. 3 results in a qualitative change to the keepers' task. In 3 vs. 2, both takers must go towards the ball because two takers are needed to capture the ball from the keeper. In 4 vs. 3, the third taker is now free to roam the field and attempt to intercept passes. This changes the optimal keeper behavior, as one teammate is often blocked from receiving a pass by a taker. Furthermore, adding a keeper in the center of the field changes the start state significantly: the keeper that starts with the ball has a teammate closer to itself but also quite close to the takers.

4.3.2 3 vs. 2 XOR Keepaway

This section describes a modification to the 3 vs. 2 Keepaway task in which the agent's internal representation must be capable of learning an "exclusive or" to achieve top performance. This serves as an example of a task where a linear learning

```
if Keeper attempts pass to closest teammate then
  if (4m < dist(K1,T1) < 6m) XOR (9m < dist(K2,T) < 12m) then
    Execute good pass
  else
    Execute bad pass
else if Keeper attempts pass to furthest teammate then
  if (9m < dist(K1,K2) < 12m) OR (45° < ang(K2) < 90°) then
    Execute good pass
  else
    Execute bad pass
else
  if Keeper could have executed good pass if it had decided to pass
  then
    Execute bad pass
  else
    Execute hold
```

Fig. 4.11 XOR Keepaway changes the effects of agent's actions but leave the rest of the task unchanged from 3 vs. 2 Keepaway.

representation can learn quickly, but is eventually outperformed by a more complex learning representation.

In XOR Keepaway, the 3 vs. 2 Keepaway task is modified to change the effect of agents' actions. Good pass executes the pass action and additionally disables the takers for 2 seconds. Bad pass causes the keeper's pass to travel directly to the closest taker. These effects are triggered based on the agent's chosen pass action and 4 state variables: the distance to the closest taker, $dist(K_1,T_1)$; the distance from the closest teammate to a taker, $dist(K_2,T)$; the passing angle to the closest teammate, $ang(K_2)$; and the distance to the closest teammate, $dist(K_1,K_2)$. Agents which lack the representational power to express an XOR can learn to improve their performance, but are unable to achieve optimal performance. This modification to the task changes the effects of the agents' decisions but leaves the rest of the task unchanged. Details appear in Figure 4.11.

4.3.3 Additional Keepaway Variants

Section 4.3.1 discussed extending 3 vs. 2 Keepaway to 4 vs. 3 Keepaway. In this monograph we will also discuss extensions of the Keepaway task to 5 vs. 4 Keepaway, 6 vs. 5 Keepaway, and 7 vs. 6 Keepaway. In these tasks, the players remain on a $25m \times 25m$ field. The start state is defined so that the extra keepers are placed near the center of the field and the takers are grouped in the same corner. In 5 vs. 4, there are 5 actions and 25 state variables. 6 vs. 5 has 6 actions and 31 state variables; 7 vs. 6 has 7 actions and 37 state variables. As discussed in the previous section, adding more players to the task increases the difficulty of the problem as well as increasing the size of the state and action spaces.

In addition to changing the number of players in Keepaway tasks, we can also change properties of the environment or properties of the players. Examples of environmental changes include field size or friction; examples of player changes include using different levels of sensor noise or actuator noise. By allowing such changes to the Keepaway task, we are able to design tasks that change in qualitatively different ways from adding players, allowing us to test transfer learning between more pairs of tasks.

In this monograph we also investigate a pair of new tasks, *3 vs. 2 Inaccurate Keepaway* and *4 vs. 3 Inaccurate Keepaway* [Taylor et al. (2007a)], created by changing the passing actuators on some sets of agents so that the passes are significantly less accurate.[10] As we will show later in Section 5.3.3, such a change causes the keepers' actions in this task to have qualitatively different transition functions from the standard Keepaway tasks.

4.3.4 *Learning Keepaway*

The Keepaway problem maps fairly directly onto the discrete-time, episodic RL framework. As a way of incorporating domain knowledge, the learners choose not from the simulator's primitive actions but from a set of higher-level macro-actions implemented as part of the player, as described by [Stone et al. (2005)]. These macro-actions can last more than one time step and the keepers have opportunities to make decisions only when an on-going macro-action terminates. The macro-actions (`Hold`, $\texttt{Pass}_1$, and $\texttt{Pass}_2$ in 3 vs. 2) that the learners select among can last more than one time step, and the keepers have opportunities to make decisions only when an on-going macro-action terminates. To handle such situations, it is convenient to treat the problem as a *semi-Markov decision process*, or SMDP [Puterman (1994), Bradtke and Duff (1995)]. The agents make decisions at discrete SMDP time steps (when macro-actions are initiated and terminated).

4.3.4.1 Learning Keepaway with Sarsa

To learn Keepaway with Sarsa, each keeper is controlled by a separate agent. CMAC, RBF, and ANN function approximation have all been successfully used to approximate an action-value function in Keepaway [Stone et al. (2005), Taylor et al. (2005)]. When using a CMAC or RBF, each state variable is tiled independently (e.g., there are 13 separate CMAC or RBF approximators in 3 vs. 2 Keepaway). All weights in the CMAC and RBF function approximators are initially set to zero; every initial state-action value is thus zero, and the action-value function is uniform. When using a feedforward ANN, we used 20 hidden units, selected via initial experimentation on six different network topologies. The number of input nodes and output nodes are set

[10] Passing actuators are changed in the benchmark players by modifying the pass action from the default "PassNormal" to "PassFast." This increases the speed of the pass by 50%, significantly reducing accuracy and causing more missed passes.

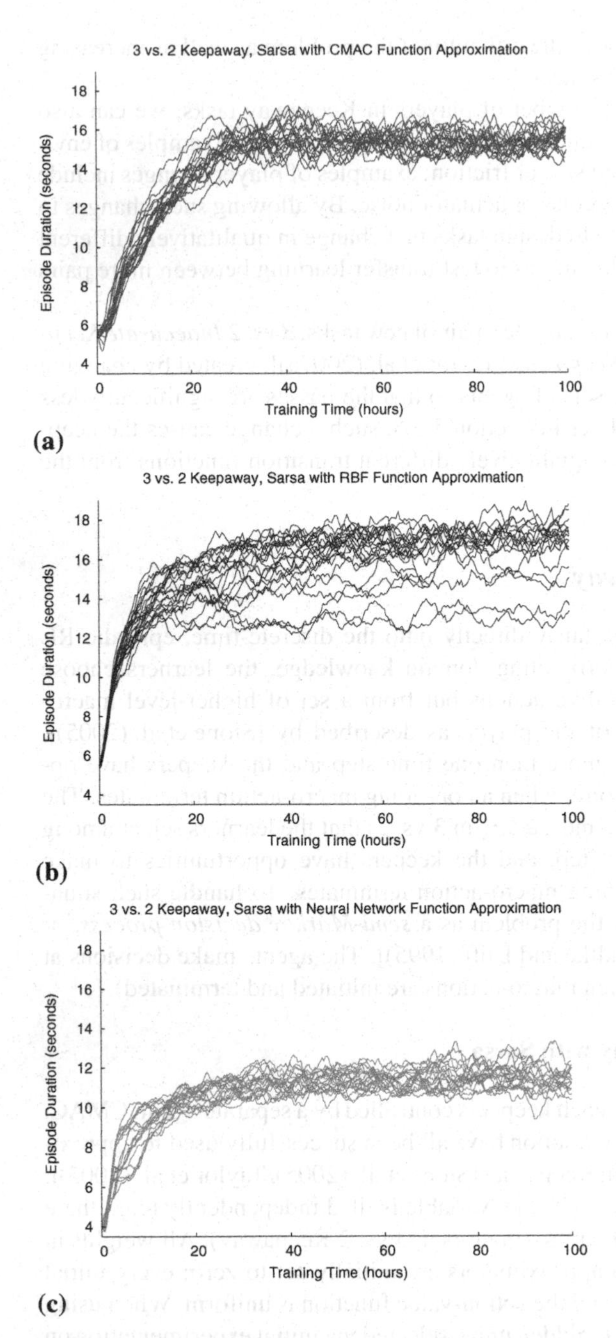

Fig. 4.12 These three graphs show qualitative learning results in 3 vs. 2 Keepaway with Sarsa on a $20m \times 20m$ field. The CMAC, RBF, and ANN function approximators are used in (a), (b), and (c), respectively. The x-axis shows the training time in simulator hours and the y-axis shows the average episode length in simulator seconds. Wall clock time is roughly half of the simulator time. These graphs use version 0.6 of the Keepaway benchmark players.

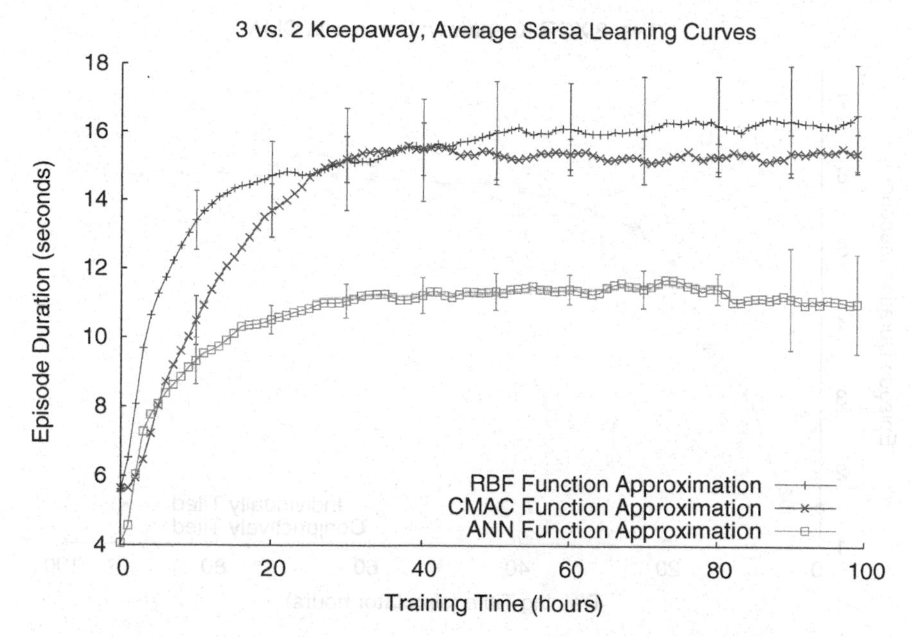

Fig. 4.13 This graph displays the average learning curves for the three function approximators (Figures 4.12a–c), along with the standard deviation

to the number of state variables and actions in the task, respectively. For example, the 3 vs. 2 task uses a 13-20-3 neural network. All weights and biases in the feed-forward ANN are given small random numbers to encourage faster backpropagation training [Mehrotra et al. (1997)], but the initial action-value is still near uniform. As training progresses, the weights of the function approximators are changed by Sarsa so that the average hold time of the keepers increases.

In our experiments we set the learning rate, α, to be 0.1 for the CMAC function approximator, as in previous experiments. α is 0.05, and 0.125 for the RBF and ANN function approximators, respectively. These values were determined after trying approximately five different learning rates for each function approximator. The exploration rate, ε, was set to 0.01 (1%) in all experiments, and λ was set to 0, which we selected to be consistent with past work [Stone et al. (2005)].

Figure 4.12 shows the online reward from 15 trials graphed with a 1,000 episode sliding window. Figure 4.13 shows the same data, but graphs average learning curves rather than individual trials. All three function approximators allow successful learning, but the neural network's total reward was the lowest. We posit that this difference is due to the ANN's *non-locality* property. When a particular CMAC weight for one state variable is updated during training, the update will affect the output value of the CMAC for other nearby state variable values. The width of the CMAC tiles determines the generalization effect, and outside of this tile width the change has no effect. Contrast this with a neural network: every weight is used for the calculation of a value function, regardless of how close two inputs are in state space. Any

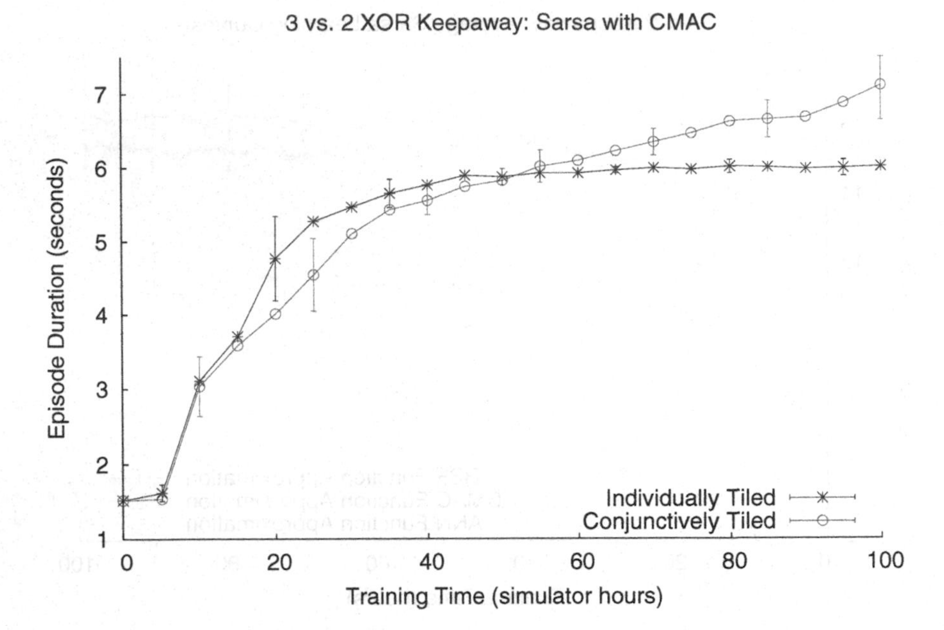

Fig. 4.14 This graph compares the performance of learning $20m \times 20m$ 3 vs. 2 XOR Keepaway using Sarsa with two different function approximations. The Individually Tiled CMAC treats each state variable independently, whereas the Conjunctively Tiled CMAC is able to learn the interdependence of different state variables.

update to a weight in the neural network must necessarily change the final output of the network for every set of inputs. Therefore it may take the neural network longer to settle into an optimal configuration. The RBF function approximator had the best average performance of the three. RBF function approximation shares CMAC's locality benefits, but is also able to generalize more smoothly due to the Gaussian summation of weights.

Figure 4.14 shows learning results on 3 vs. 2 XOR Keepaway with Sarsa learners. The players labeled "Individually Tiled CMAC" treat each state variable independently, while the players labeled "Conjunctively Tiled CMAC" conjunctively tile the four critical state variables (as described in Figure 4.11) together in a 4-dimensional CMAC, treating the remaining 9 state variables independently. The simpler representation that only considers each state variable in isolation is able to achieve higher performance than the more complex representation initially, but ultimately plateaus at a lower performance level. Section 7.3.4 will present experiments showing that it is possible to transfer from one representation to another and thus receive the benefits of both learning representations. Regardless of the representation, however, the XOR task is significantly harder than 3 vs. 2 Keepaway, as can be seen in the decreased performance relative to the standard 3 vs. 2 Keepaway task.

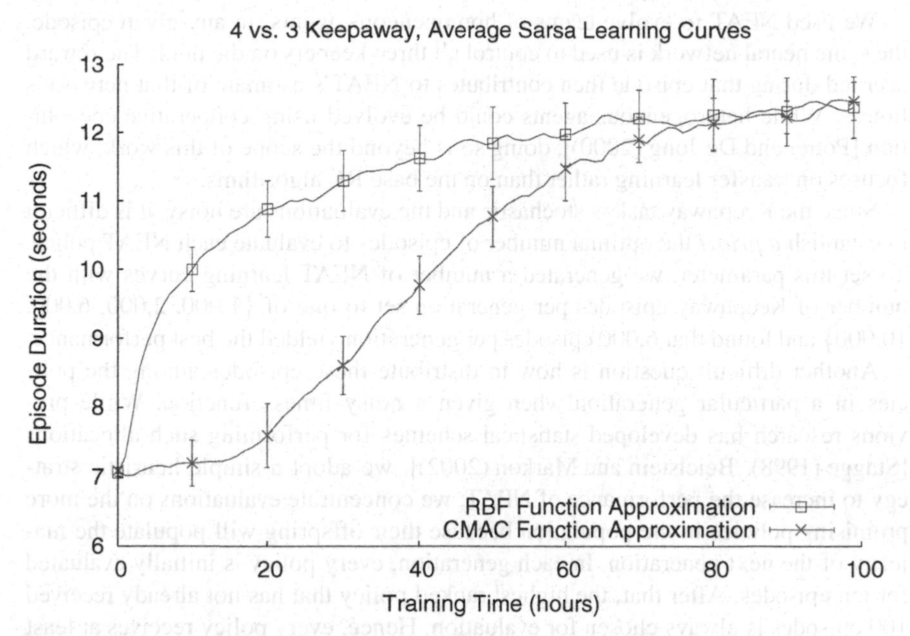

Fig. 4.15 This graph shows that RBF function approximation is superior to CMAC function approximation on $25m \times 25m$ 4 vs. 3 Keepaway. 14 trails are averaged for each setting. Error bars show the standard deviation of the two methods.

Figure 4.15 shows the performance of a Sarsa learner using CMAC function approximation and RBF function approximation in 4 vs. 3 Keepaway. As in 3 vs. 2 Keepaway, RBF learners are significantly faster than CMAC learners.

The learning setup for 3 vs. 2 Inaccurate Keepaway and 4 vs. 3 Inaccurate Keepaway is similar to the accurate versions of the tasks. As discussed later in Section 5.3.3, keepers require more training to reach a threshold performance level when their passing actuators are inaccurate, because the task is significantly harder.

4.3.4.2 Learning Keepaway with NEAT

Every network evolved by NEAT for 3 vs. 2 Keepaway has 13 inputs, corresponding to the Keepaway state variables, and 3 outputs, corresponding to the available macro-actions. The keepers always select the action with the highest activation, breaking ties randomly. We found that a population of 100 policies with a target of 5 species and the default values of $c_1 = 1.0$, $c_2 = 1.0$, and $c_3 = 2.0$ allowed NEAT to learn well in the Keepaway task [Taylor et al. (2006)].[11]

[11] [Stanley and Miikkulainen (2002)] describe the semantics of the NEAT parameters in detail. When using NEAT to learn Keepaway tasks, we used the following additional parameters. The compatibility distance δ_t was adjusted dynamically to maintain a target of 5 species. The survival threshold was 0.2, the weight mutation power was 0.01, the interspecies mating rate was 0.05, the drop-off age was 1,000, and the probability of adding recurrent links was 0.2.

We used NEAT to evolve teams of homogeneous agents: in any given episode, the same neural network is used to control all three keepers on the field. The reward accrued during that episode then contributes to NEAT's estimate of that network's fitness. While heterogeneous agents could be evolved using cooperative coevolution [Potter and De Jong (2000)], doing so is beyond the scope of this work, which focuses on transfer learning rather than on the base RL algorithms.

Since the Keepaway task is stochastic and the evaluations are noisy, it is difficult to establish *a priori* the optimal number of episodes to evaluate each NEAT policy. To set this parameter, we generated a number of NEAT learning curves with the number of Keepaway episodes per generation set to one of {1,000, 2,000, 6,000, 10,000} and found that 6,000 episodes per generation yielded the best performance.

Another difficult question is how to distribute these episodes among the policies in a particular generation when given a noisy fitness function. While previous research has developed statistical schemes for performing such allocations [Stagge (1998), Beielstein and Markon (2002)], we adopt a simple heuristic strategy to increase the performance of NEAT: we concentrate evaluations on the more promising policies in the population because their offspring will populate the majority of the next generation. In each generation, every policy is initially evaluated for ten episodes. After that, the highest ranked policy that has not already received 100 episodes is always chosen for evaluation. Hence, every policy receives at least

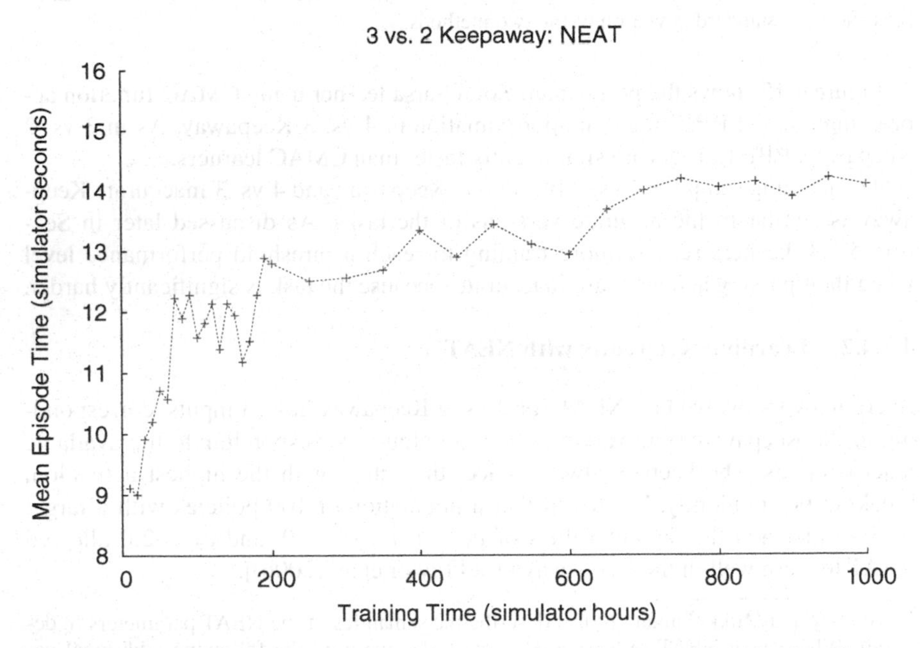

Fig. 4.16 This graph shows a learning curve averaged over 5 NEAT trials. After learning has ended, the champion from each generation is tested for 1,000 episodes, and these off-line evaluations are used to plot the agents' performance. This graph was generated with the 0.5 version of the Keepaway benchmark agents.

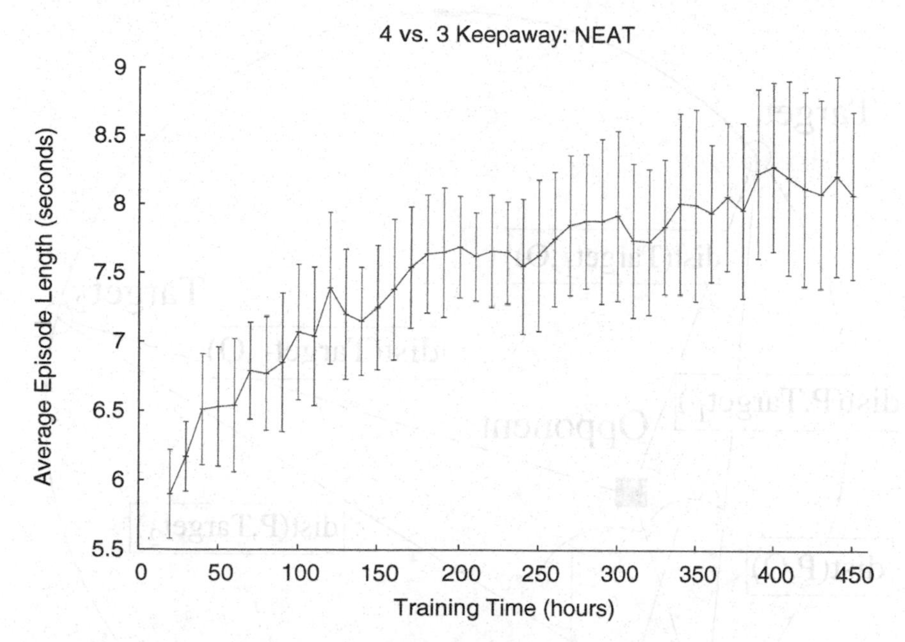

Fig. 4.17 This graph shows a learning curve averaged over 10 NEAT trials. After learning has ended, the champion from each generation is tested for 1,000 episodes, and these off-line evaluations are used to plot the agents' performance. Error bars show the standard deviation. This graph was generated with the 0.5 version of the Keepaway benchmark agents.

10 evaluations and no more than 100, with the more promising policies receiving the most.

NEAT learns significantly slower than Sarsa, but may achieve a higher asymptotic performance on some variants of 3 vs. 2 Keepaway [Taylor et al. (2006)]. Figures 4.16 and 4.17 shows example learning curves in 3 vs. 2 and 4 vs. 3; NEAT agents take roughly an order of magnitude more time to converge when compared to Sarsa agents.

4.4 Ringworld

Having introduced the Keepaway domain in the previous section, this section and the next introduce two novel tasks in the gridworld domain which were designed to be similar to 3 vs. 2 Keepaway. We are interested in transfer not only between tasks within a single domain, but also between tasks in different domains (i.e., *cross-domain transfer*). In order to show that cross-domain transfer is feasible, we construct the *Ringworld* [Taylor and Stone (2007a)] task, which can be used as a source task to speed up learning in 3 vs. 2 Keepaway.

The state space of Ringworld is discretized into $0.01m^2$ tiles, there is no noise in agents' perceptions, and only one agent (the player) is learning. This is in contrast

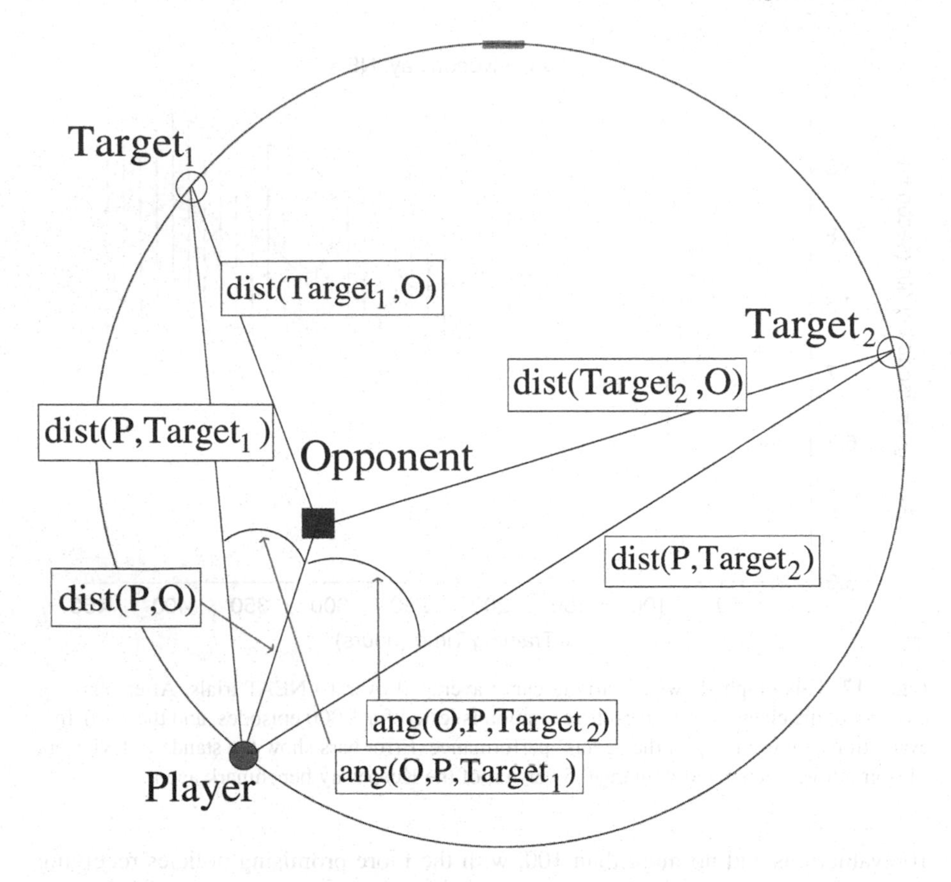

Fig. 4.18 The Ringworld player may stand still or run to 1 of 2 possible target locations. The episode ends when the opponent captures the player. The fixed $0.01m$ grid used to discretize the state space is shown for the top of the ring and shown in detail in Figure 4.19.

with Keepaway, which has a continuous state space, noisy perceptions, and is a multi-agent learning problem.

The goal of a Ringworld player is to avoid capture by an opponent for as long as possible, where capture is possible if the distance between the opponent and player is less than $1m$. The agent receives a reward of +1 for every time step in which it is not captured by the opponent. The opponent always moves towards the player. When the episode starts, the player is randomly assigned two possible "Run Targets," which always lie on a fixed ring (see Figure 4.18). At each timestep, the player may either `stay` in its current location, or choose to `run` to one of the two targets. If the player runs, it moves at twice the speed of the opponent directly towards the chosen run target. If the opponent does not intercept the player, as determined by the transition function, two new random run targets on the ring are chosen for the player and the episode continues. As the opponent approaches the player, either when the player is standing still or while running, the chance of capture increases. The only

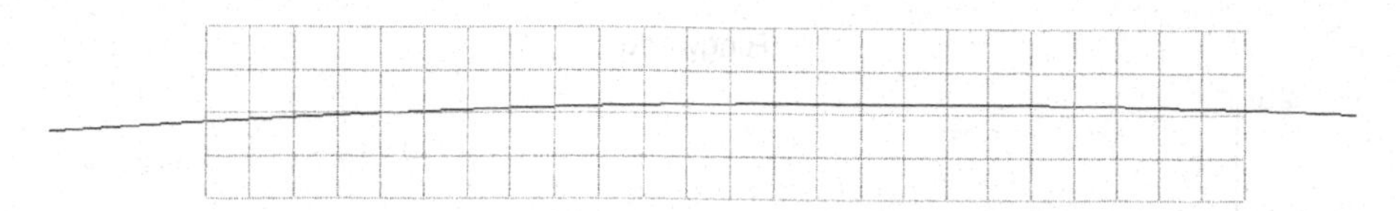

Fig. 4.19 This figure shows the detail of the underlying $0.01m$ grid near the top of the ring from Figure 4.18

Table 4.6 This table summarizes the state space and the action space of Ringworld

	State Variables	
Ringworld	*dist(P, O)*	Distance from player to opponent
	dist(P, Target$_1$)	Distance from player to near location
	dist(P, Target$_2$)	Distance from player to far location
	dist(Target$_1$, O)	Distance from near location to opponent
	dist(Target$_2$, O)	Distance from far location to opponent
	ang(O, P, Target$_1$)	Open angle between opponent and near location
	ang(O, P, Target$_2$)	Open angle between opponent and far location
	Actions	
	`Stay`	Do not move
	`Run`$_{Near}$	Run to the near location
	`Run`$_{Far}$	Run to the far location

stochasticity in the environment is the randomness associated with the probability of capture. The state is represented by 5 distances and 2 angles (see Table 4.6).

This gridworld task was constructed to have similarities to 3 vs. 2 Keepaway; for instance, the 7 state variables were chosen to be similar to the state variables in Keepaway. The width of the ring ($9.5m$) was selected so that the distance between runs is similar, on average, to the distance between keepers when playing 3 vs. 2 Keepaway. The Ringworld transition function, T, takes as input the state variable *dist(P,O)* and determines if the opponent captures the player, ending the episode. This function was constructed using the observed likelihood of whether a Keepaway episode ends, given *dist(K_1,T_1)*. While it is impossible to recreate many of the dynamics associated with a complex, stochastic, and continuous task, Ringworld captures some of Keepaway's characteristics. We show later (see Section 7.2) that we can effectively transfer between Ringworld and Keepaway although the tasks are quite different.

4.4.1 Learning Ringworld

We learn Ringworld using Sarsa with tabular function approximation. Because Ringworld is a relatively simple task, episodes can be run orders of magnitude faster than Keepaway. We found that an agent sees an average of 8,100 distinct states over the course of a 25,000 episode learning trial, taking roughly 3 minutes of wall clock time

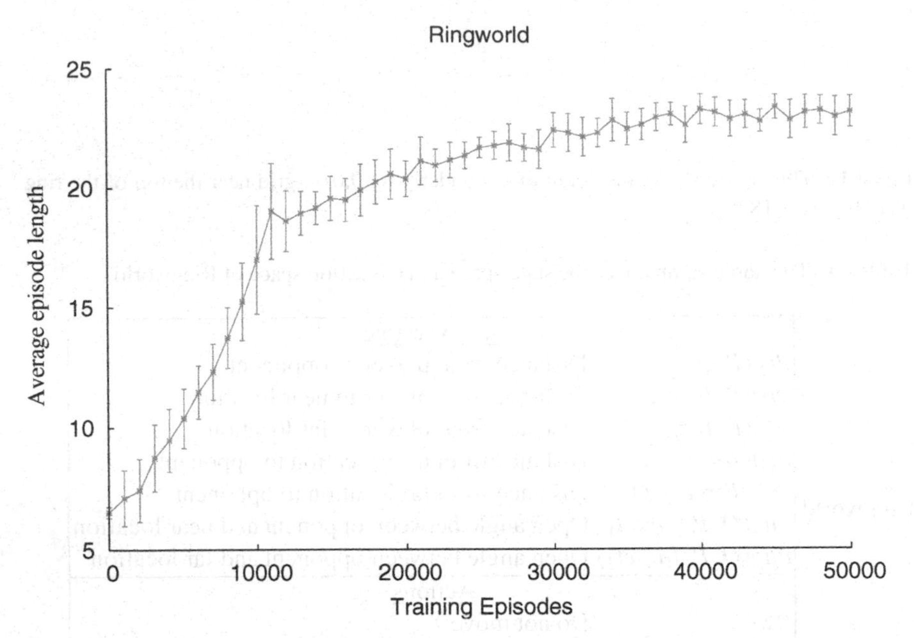

Fig. 4.20 Learning with tabular function approximation takes many episodes, but each trial is orders of magnitude faster than Keepaway in terms of wall clock time. This graph shows the average on-line performance of 10 learning trials with standard error bars.

on an Intel 3.4 MHz desktop machine. A representative learning curve is shown in Figure 4.20.

4.5 Knight Joust

Knight Joust [Taylor and Stone (2007a)] is also situated in the gridworld domain. It was designed to be less similar to Keepaway than Ringworld is, but still retain enough commonalities to enable successful transfer into Keepaway. In this task the player begins on one end of a 25 × 25 board, the opponent begins on the other end, and the players alternate moves. The player's goal is to reach the opposite end of the board without being touched by the opponent (see Figure 4.21); the episode ends if the player reaches the goal line or the opponent is on the same square as the player. The state space is discretized into squares ($1m^2$ each) and there is no noise in the perception. The player's state variables are composed of the distance from the player to the opponent, and two angles which describe how much of the goal line is viewable by the player.

The player chooses between deterministically moving one square North or performing a *knight's jump*, moving one square North and then two East or West: A = {`Forward`, `Jump`$_{West}$, `Jump`$_{East}$}. Table 4.7 lists the state variables and actions for Knight Joust. The opponent may move in any of 8 directions and follows a fixed stochastic policy. If the opponent had an optimal policy, the player could never

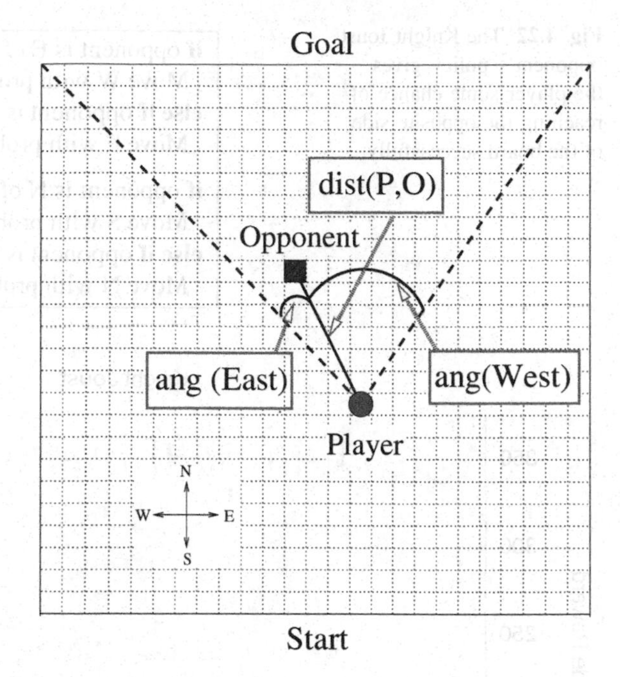

Fig. 4.21 Knight Joust: The player attempts to reach the goal end of a 25 × 25 gridworld and the opponent attempts to tag the player

Table 4.7 This table summarizes the state space and the action space of the Knight Joust task

Knight Joust		State Variables
	dist(P, O)	Distance from player to opponent
	ang(West)	Open angle to West
	ang(East)	Open angle to East
		Actions
	`North`	Move one square North
	$\texttt{Jump}_{West}$	Move one square North and two West
	$\texttt{Jump}_{East}$	Move one square North and two East

successfully reach the goal line; the player can only take a limited number of jumps before hitting the edge of the board. The player must instead rely on the opponent to "stumble" so that the player can pass it. The opponent's policy is summarized Figure 4.22.

The player receives a reward of +5 every time it takes the forward action, a +50 upon reaching the goal line, and 0 otherwise.[12] While this task is quite dissimilar from Keepaway (Knight Joust is fully observable, has a discrete state space, the player's actions are deterministic, and only a single agent learns), note that there are some similarities, such as favoring larger distances between player and opponent.

[12] In one set of experiments in Section 7.2.2.2, we use an older version of this task where the player receives a reward of +20 for every forward actions and a reward of +20 for reaching the goal line.

Fig. 4.22 The Knight Joust opponent's policy gives the player some chance of reaching the opposite side of the board successfully.

if opponent is E of player **then**
  Move W with probability 0.9
else if opponent is W of player **then**
  Move E with probability 0.9

if opponent is N of player **then**
  Move S with probability 1.0
else if opponent is S of player **then**
  Move N with probability 0.8

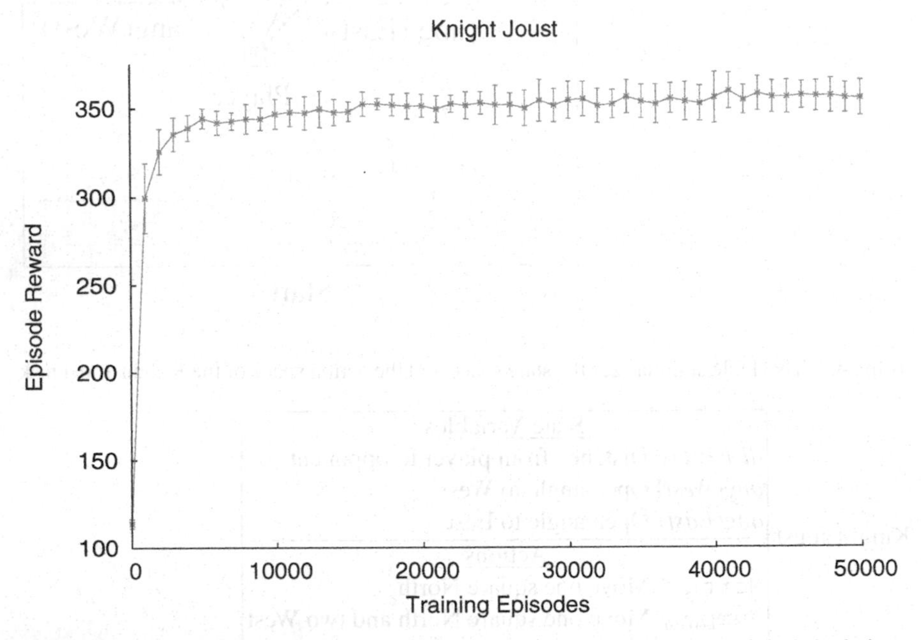

Fig. 4.23 Each 50,000 episode trail of Knight Joust takes less than a minute of wall clock time on an Intel 3.4 MHz desktop machine. This graph shows the average on-line performance of 30 learning trials with standard error bars.

We will show in Sections 5.3.4 and 7.2 that our transfer algorithms can effectively improve learning in 3 vs. 2 Keepaway after first learning in Knight Joust.

4.5.1 Learning Knight Joust

Knight Joust, like Ringworld, can be learned with Sarsa using a tabular representation. The agent experiences an average of only 600 distinct states in a 50,000 episode learning trial. Figure 4.23 shows an average learning curve.

Players learning 3 vs. 2 Keepaway takes roughly 20 hours of wall-clock time to achieve their asymptotic performance (which is equivalent to 40 simulator hours in the RoboCup Soccer Server), but Ringworld players require only 3 minutes of wall clock time. Knight Joust requires even less time. Such differences in training time highlight the possible benefit of cross-domain transfer. As we show later (Chapters 5.3.4 and 7.2.2), even though Ringworld and Knight Joust are much simpler than Keepaway, they can be successfully used as source tasks for transfer into Keepaway. If a researcher was attempting to minimize the wall-clock total training time, using a gridworld task as a source is attractive because agents can train in a negligible amount of time compared to learning a Keepaway task.

4.6 Summary of Domains

In this chapter we have discussed a number of tasks in four different domains: Mountain Car, Server Job Scheduling, Keepaway, and gridworld. These tasks were chosen so that the experiments in later chapters demonstrate the flexibility of transfer: in addition to using different learning methods and function approximators, utilizing tasks with different characteristics shows that our TL algorithms function in

Table 4.8 This table summarizes characteristics of domains used in this monograph. Tasks marked with a † are novel.

Selected Domain and Task Characteristics

Domain	Task	# Actions	# State Variables	Discrete State Space?	Deterministic Transition Function?	Fully Observable?
Mountain Car	Standard 2D	3	2	N	Y	Y
	No Goal 2D†	3	2	N	Y	Y
	Hand Brake 2D†	4	2	N	Y	Y
	Low Power 2D†	3	2	N	Y	Y
	High Power 2D†	3	2	N	Y	Y
	Standard 3D†	5	4	N	Y	Y
	Lower Power 3D†	5	4	N	Y	Y
	Hand Brake 3D†	6	4	N	Y	Y
Server Job Scheduling	2-job-type	8	8	Y	N	Y
	4-job type	16	16	Y	N	Y
Keepaway	3 vs. 2	3	13	N	N	N
	4 vs. 3	4	19	N	N	N
	5 vs. 4	5	25	N	N	N
	6 vs. 5†	6	31	N	N	N
	7 vs. 6†	7	37	N	N	N
	3 vs. 2 XOR†	3	13	N	N	N
	3 vs. 2 Inaccurate†	3	13	N	N	N
	4 vs. 3 Inaccurate†	4	19	N	N	N
Gridworld	Ringworld†	3	7	Y	N	Y
	Knight Joust†	3	3	Y	N	Y

a wide variety of settings. The tasks and their most salient characteristics are summarized in Table 4.8.

The existing tasks discussed in this chapter and the novel tasks introduced by this chapter serve two purposes. We selected tasks from the Server Job Scheduling domain and the Keepaway domain that existed in the literature; the fact that we were able to readily find tasks suitable for transfer is a good indication that our TL methods are applicable to existing problems.

The novel tasks were introduced to test the applicability of our TL methods and examine particular questions. For instance, in the Mountain Car domain, Standard 2D and Standard 3D are more similar than High Power 2D and Standard 3D, and thus we expect the former pair to be more successful for transfer than the later pair. The XOR Keepaway task was introduced to serve as an existence proof of a task that could be learned with multiple representations: a simple representation which could learning quickly, and a more complex representation which could learn to achieve better performance at the expense of requiring more data.

The next chapter details a key concept in this monograph, Value Function Transfer, the first of six TL methods detailed in this monograph capable of successful transfer via inter-task mappings.

Chapter 5
Value Function Transfer via Inter-Task Mappings

The previous chapter introduced a number of different tasks and demonstrated that all could be learned with existing RL techniques. This chapter presents the core contributions of the monograph: *inter-task mappings* [Taylor et al. (2007a)], *Value Function Transfer* [Taylor et al. (2007a)], and empirical results demonstrating their effectiveness in the Keepaway domain.

Transfer methods in this monograph focus on pairs of tasks with different state variables and actions. For a given pair of tasks which we wish to transfer between, we use the notation S_{source} and A_{source} to denote the states and actions for the source task, and S_{target} and A_{target} for states and actions in the target task.

The next section introduces inter-task mappings, a novel construct that enables transfer between tasks with different state variables and actions. This monograph uses such mappings to transfer action-value functions, policies, rules, and instances between source and target tasks. Section 5.2 introduces Value Function Transfer, a novel TL method that utilizes inter-task mappings to effectively transfer a learned action-value function from a source task to a target task. Section 5.3 presents results from the Keepaway and Knight Joust domains showing that Value Function Transfer can significantly increase learning performance in a variety of tasks.

The following two chapters will present additional TL methods and empirical results that make use of inter-task mappings. In Chapters 5–7, we assume that all inter-task mappings are provided to the agent and are correct. In Chapter 8, we relax this assumption and introduce two methods that can autonomously learn such mappings.

5.1 Inter-Task Mappings

In order to effectively transfer between tasks that have different state variables and actions, some type of mapping is typically necessary (as introduced in Section 3.1.3). In addition to knowing that a source task and target task are related, a TL method often needs to know *how* they are related. Inter-task mappings provide such a relation.

M.E. Taylor: Transfer in Reinforcement Learning Domains, SCI 216, pp. 91–120.
springerlink.com

Consider an agent that is presented with a target task that has a set of actions (A_{target}). If the agent knows how those actions are related to the action set in the source task (A_{source}), it is much more likely that it will be able to effectively reuse knowledge gained in the source task. (For the sake of exposition we focus on actions, but an analogous argument holds for state variables.) The majority of transfer algorithms assume that no explicit task mappings are necessary because the source and target task have the same state variables and actions. However, if this assumption is false, the agent needs to be told, or learn, how the two tasks are related.

In addition to having the same labels, the state variables and actions need to have the same semantic meanings in both tasks if no mapping is used. For instance, consider the Mountain Car domain. Suppose that the source task had the actions A = {`Forward, Neutral, Backward`}. If the target task had the actions A = {`Right, Neutral, Left`}, a TL method would need some kind of mapping because the actions had different labels. Furthermore, suppose that the target task had the same actions as the source (A = {`Forward, Neutral, Backward`}) but the car was facing the opposite direction, so that `Forward` accelerated the car in the negative x direction and `Backward` accelerated the car in the positive x direction. If the cardinality of two action sets is not equal, some actions may have no equivalence in the two tasks. Mappings would be less critical if the agent had access to semantic information about the tasks, but such information is typically not provided to agents in the RL framework and this monograph does not assume that such information is available.

We define an *action mapping* (χ_A) that maps actions in the target task to actions in the source task that have "similar" effects on the environment. Similarity depends on how the transition and reward functions in the two MDPs are related. Recall that figure 3.1 depicts an action mapping as well as a *state-variable mapping* (χ_X) between two tasks. Transfer scenarios considered in this monograph typically have more actions and state variables in the target task than in the source task; a source task is thus used to learn a relatively complex target task faster than if transfer was not used. If the two tasks had the same number of actions, the action mapping could be one-to-one. For instance, one could define χ_A(`Neutral`) = `Neutral`, χ_A(`East`) = `Right` and χ_A(`West`) = `Left` in the previous example. If the source task instead had more actions than the target task, we could construct a one-to-one action mapping that ignores the irrelevant source task actions, or construct a one-to-many mapping. Which type of mapping to choose, and its construction, depends on the particulars of the tasks in question.[1] In Section 5.1.1 we provide examples of action and state variable mappings using tasks from the Keepaway domain.

Rather than defining full mappings, in some situations *partial inter-task mappings* [Taylor et al. (2007b)] may be more appropriate, depending on the amount of knowledge available about the source task and the target task. In a partial mapping, novel actions in the target task are ignored. Continuing the above example, consider an alternative target task that included a third action, `North`. Suppose that the

[1] This monograph focuses on many-to-one mappings under the assumption that the target task is larger than the source task. However, TL methods in this monograph should be applicable to one-to-many mappings as well.

human creating the mappings (or an agent learning the mappings) could not determine if the target task action `North` was similar to either of the source task actions, `Left`, `Neutral`, or `Right`. The partial mapping could again map `Neutral` to `Neutral`, `West` to `Left`, and `East` to `Right`, but not map `North` to any source task action. Such a partial mapping may be easier to intuit than full mappings, or may be all that's justifiable, particularly if the target task's actions are a superset of the source task's actions (this is the case in the Keepaway domain, as discussed in Section 5.1.1).

For a given pair of tasks, there could be many ways to formulate inter-task mappings. If a human is in the loop, assisting the agent as it learns, a TL method may be provided with inter-task mappings. However, if the agent is expected to transfer autonomously, such mappings have to be learned. In this case, learning methods attempt to minimize the amount of samples needed, and/or the computational complexity of the learning method, while still learning a mapping that enables effective transfer. This monograph introduces two methods in Chapter 8 that are able to learn such mappings off-line from experience gathered in a pair of tasks, and existing learning methods for inter-task mappings will be discussed as part of the related work (Chapter 3). All TL methods presented in this monograph are capable of using inter-task mappings, regardless of whether they are provided or learned autonomously.

5.1.1 Inter-Task Mappings for the Keepaway Domain

In this section we introduce hand-coded inter-task mappings for tasks in the Keepaway domain. In addition to detailing mappings which will be used in later experiments in this monograph, this section grounds out the discussion of inter-task mappings in a set of examples.

One advantage of using the Keepaway domain for TL research is that inter-task mappings between different tasks may be easily generated. In general, mappings between tasks may not be so straightforward, but experimenting in a domain where they are easily defined allows us to focus on showing the benefits of transfer. Experimental results using these mappings are presented in Section 5.3.

To construct an inter-task mapping between 4 vs. 3 Keepaway and 3 vs. 2 Keepaway, we first define χ_A by identifying actions that have similar effects on the environment in both tasks. For the 3 vs. 2 and 4 vs. 3 tasks, the action `Hold ball` is considered to be equivalent; this action has a similar effect on the environment in both tasks. Likewise, the action labeled `Pass to closest keeper` is analogous in both tasks, as is `Pass to second closest keeper`. We choose to map the novel target action, `Pass to third closest keeper`, to `Pass to second closest keeper` in the source task.

The state variable mapping, χ_X, is constructed using a similar strategy. Each of the 19 state variables in the 4 vs. 3 task is mapped to a similar state variable in the 3 vs. 2 task. For instance, the state variable labeled *Distance to closest keeper* is the same in both tasks. The *Distance to second closest keeper* state variable in the target task is similar to *Distance to second closest keeper* state variable in the source task. *Distance to third closest keeper* in the target task is mapped to *Distance to second closest keeper*, which is the most similar state variable in the source task. See Table 5.1 for a full description of χ_X and χ_A.

We do not claim that this definition of χ_A and χ_X is optimal for the Keepaway domain. However, experiments show that it does enable effective transfer. Further investigation demonstrates that possible alternate inter-task mappings are less effective for transfer than χ_A and χ_X (see Section 5.3.2).

The information to relate the two tasks, as contained in χ_X and χ_A, may not always be available. For example, when the target task is an extension of the source task, we may know which state variables and actions are most similar in the two tasks but not know how to map *novel* state variables and actions back to the source task. In Keepaway, this would mean that the human (or learning algorithm) identified which actions and state variables in 4 vs. 3 Keepaway were related to keepers 1–3 and takers 1–2, but did not know how to handle actions and state variables having to do with the fourth keeper or third taker. Thus χ_A would map the 4 vs. 3 actions `Hold ball`, `Pass to closest keeper`, and `Pass to second closest keeper` to their equivalent actions in 3 vs. 2, but leave χ_A(`Pass to third closest keeper`) undefined. We will see later (Sections 5.3.2 and 6.2.3) that this partial mapping allows transfer to significantly outperform learning without transfer on the 4 vs. 3 task, but it is inferior to the fully defined inter-task mapping.

χ_A for 5 vs. 4 to 4 vs. 3 can be constructed analogously to the full mapping between 4 vs. 3 and 3 vs. 2. The novel action in 5 vs. 4 is mapped to `Pass to the third keeper` in the source task, and the novel state variables are mapped to the fourth keeper and third taker in the source task. Specifically, the target task actions `Hold ball`, `Pass to closest keeper`, `Pass to second closest keeper`, and `Pass to third closest keeper` are all mapped to the corresponding actions in 4 vs. 3. The novel 5 vs. 4 action, `Pass to fourth closest keeper`, is mapped to the 4 vs. 3 action `Pass to third closest keeper`. χ_X then follows the same pattern – novel state variables are mapped to the most similar state variables in the 4 vs. 3 task.

All mappings from n vs. m Keepaway to $(n-1)$ vs. $(m-1)$ Keepaway may be constructed following this pattern. TL experiments that utilize mappings for 6 vs. 5 and 7 vs. 6 are presented in Section 5.3.6.

We additionally construct χ_A for 5 vs. 4 to 3 vs. 2. Now the actions `Hold ball`, `Pass to closest keeper`, and `Pass to second closest keeper` are mapped from 5 vs. 4 to the same titled actions in 3 vs. 2. However, both 5 vs. 4 actions `Pass to third closest keeper` and `Pass to fourth closest keeper` are mapped to the 3 vs. 2 action `Pass to second closest keeper`. χ_X is created analogously – state variables for the two novel keepers and two novel takers are mapped to state variables for the third keeper and second taker in 3 vs. 2. Both 3 vs. 2 and 4 vs. 3 may therefore be used as source tasks for 5 vs. 4 Keepaway. We will show in Section 5.3.6 that both inter-task mappings enable successful transfer, but that a two-step transfer, from 3 vs. 2 into 4 vs. 3 and then to 5 vs. 4, is superior to a single step transfer between 3 vs. 2 and 5 vs. 4.

Table 5.1 This table describes the mapping between state variables in 4 vs. 3 Keepaway and state variables in 3 vs. 2 Keepaway. Distances and angles not present in 3 vs. 2 are in bold.

Description of χ_X Mapping from 4 vs. 3 to 3 vs. 2

4 vs. 3 state variable	3 vs. 2 state variable
$dist(K_1,C)$	$dist(K_1,C)$
$dist(K_1,K_2)$	$dist(K_1,K_2)$
$dist(K_1,K_3)$	$dist(K_1,K_3)$
$\mathbf{dist(K_1,K_4)}$	$dist(K_1,K_3)$
$dist(K_1,T_1)$	$dist(K_1,T_1)$
$dist(K_1,T_2)$	$dist(K_1,T_2)$
$\mathbf{dist(K_1,T_3)}$	$dist(K_1,T_2)$
$dist(K_2,C)$	$dist(K_2,C)$
$dist(K_3,C)$	$dist(K_3,C)$
$\mathbf{dist(K_4,C)}$	$dist(K_3,C)$
$dist(T_1,C)$	$dist(T_1,C)$
$dist(T_2,C)$	$dist(T_2,C)$
$\mathbf{dist(T_3,C)}$	$dist(T_2,C)$
$\mathrm{Min}(dist(K_2,T_1), dist(K_2,T_2), \mathbf{dist(K_2,T_3)})$	$\mathrm{Min}(dist(K_2,T_1), dist(K_2,T_2))$
$\mathrm{Min}(dist(K_3,T_1), dist(K_3,T_2), \mathbf{dist(K_3,T_3)})$	$\mathrm{Min}(dist(K_3,T_1), dist(K_3,T_2))$
$\mathrm{Min}(\mathbf{dist(K_4,T_1)}, \mathbf{dist(K_4,T_2)}, \mathbf{dist(K_4,T_3)})$	$\mathrm{Min}(dist(K_3,T_1), dist(K_3,T_2))$
$\mathrm{Min}(ang(K_2,K_1,T_1), ang(K_2,K_1,T_2), \mathbf{ang(K_2,K_1,T_3)})$	$\mathrm{Min}(ang(K_2,K_1,T_1), ang(K_2,K_1,T_2))$
$\mathrm{Min}(ang(K_3,K_1,T_1), ang(K_3,K_1,T_2), \mathbf{ang(K_3,K_1,T_3)})$	$\mathrm{Min}(ang(K_3,K_1,T_1), ang(K_3,K_1,T_2))$
$\mathrm{Min}(\mathbf{ang(K_4,K_1,T_1)}, \mathbf{ang(K_4,K_1,T_2)}, \mathbf{ang(K_4,K_1,T_3)})$	$\mathrm{Min}(ang(K_3,K_1,T_1), ang(K_3,K_1,T_2))$

5.2 Value Function Transfer

The previous section defined inter-task mappings and gave examples from the Keepaway domain. This section introduces Value Function Transfer, the core transfer learning method in this monograph.

The goal of Value Function Transfer is to use the learned action-value function from the source task, $Q_{(source,final)}$, to initialize the action-value function of a TD learner in a related target task. However, the learned action-value function cannot simply be copied over directly because it may not be defined on the target task's state variables and actions.

Value Function Transfer relies on a transfer functional, $\rho(Q)$, which allows an agent to use a source task learner's Q-function to initialize a target task learner's Q-function (see Figure 5.1). The transfer functional ρ needs to modify the action-value function so that it accepts S_{target} and A_{target} as inputs. This transformed action-value function may not provide immediate improvement over acting randomly in the target task (as measured by the jumpstart TL metric), but it should bias the

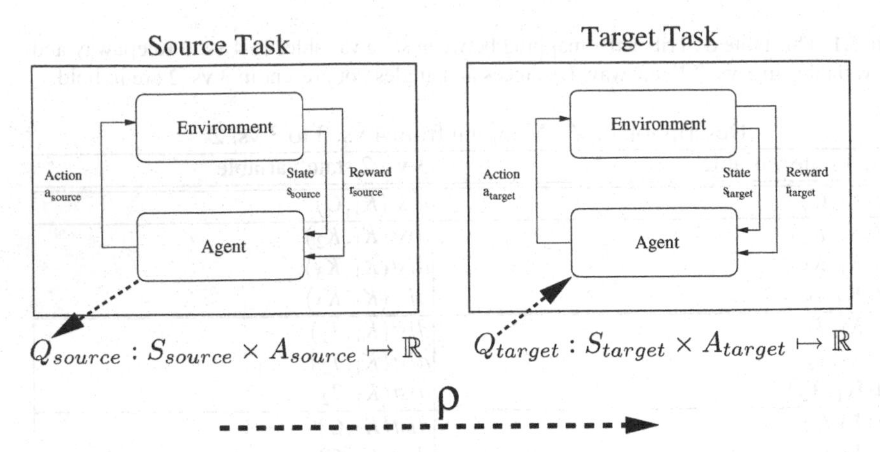

Fig. 5.1 ρ is a functional that transforms a state-action function Q from a source task so that it is applicable in a target task with different state and action spaces

learner so that it is able to learn the target task faster than if it were learning without transfer. Defining ρ to do so correctly is the key technical challenge for this TL method. The construction of ρ depends on the type of function approximator used by the agent. However, for each type of function approximator, inter-task mappings can be automatically used by a function-approximator-specific ρ to enable transfer between different source and target tasks. We introduce ρ_{CMAC}, ρ_{RBF}, and ρ_{ANN} in the next two sections which can be used with agents utilizing CMAC, RBF, and ANN function approximation, respectively.

It may seem counterintuitive that this TL method uses low-level action-value function information to speed up learning across different tasks, rather than attempting to abstract knowledge so that it is applicable to more general tasks. For instance, an agent could be trained to balance a pole on a cart and then be asked to balance a pair of poles on a cart. An example of abstract knowledge in this domain would be things like "avoid hitting the end of the track," and "it is better to have the pole near vertical." Instead of trying to transfer higher level information about a source task into a target task, Value Function Transfer instead focuses on information contained in individual weights within function approximators (or information about individual state, action pairs, in the case of a tabular action-value function). In this example, such weights would contain specific information such as how fast to move the cart to the left when a pole was at a particular angle. Weights that encode this type of low level knowledge are the most task-specific part of the learner's knowledge, but it is exactly these task-dependant details that allow Value Function Transfer to achieve significant speedups on similar tasks.

5.2.1 Constructing ρ_{CMAC} and ρ_{RBF}

As discussed in Section 2.2.1, the CMAC function approximator is trained via temporal difference learning so that its input is a (state, action) pair and its output is

Algorithm 6. APPLICATION OF ρ_{CMAC}

1: **for** each non-zero weight, w_i in the source CMAC **do**
2: $x_{source} \leftarrow$ state variable corresponding to tile i
3: $a_{source} \leftarrow$ action corresponding to tile i
4: **for** each value x_{target} such that $\chi_X(x_{target}) = x_{source}$ **do**
5: **for** each value a_{target} such that $\chi_A(a_{target}) = a_{source}$ **do**
6: $j \leftarrow$ the tile in the target CMAC activated by x_{target}, a_{target}
7: $w_j \leftarrow w_i$
8: $w_{average} \leftarrow$ average value of all non-zero weights in the target CMAC
9: **for** each weight w_j in the target CMAC **do**
10: **if** $w_j = 0$ **then**
11: $w_j \leftarrow w_{average}$

the expected return (discounted) reward. ρ_{CMAC} is a transfer functional that can be applied to CMAC function approximators. ρ_{CMAC} is used by Value Function Transfer so that when the target task learner evaluates a novel (s_{target}, a_{target}) pair, the weights for the activated tiles in the target task CMAC are not zero, but instead have been initialized by $Q_{(source,final)}$. To accomplish this, after learning the source task but before learning begins in the target task, Value Function Transfer copies weights learned in the source CMAC into weights in a newly initialized target CMAC. Algorithm 6 describes the transfer functional and shows how χ_X and χ_A are leveraged (lines 4–7).

As a final step (Algorithm 6, lines 8–11), any weights which have not yet been initialized in the target task CMAC are set to the average value of all initialized weights. When transferring between Keepaway tasks, it is often the case that the source task training was not exhaustive: some weights which may be utilized in 4 vs. 3 would otherwise remain uninitialized if not for this final step. By setting these weights to an average value, every weight in the target task CMAC can be initialized to values which have been determined via training in 3 vs. 2. This averaging effect is discussed further in Section 5.3.2, along with other possible options, and can be considered a heuristic that allows agents in the target task to learn faster in practice.

If a 4 vs. 3 keeper uses Value Function Transfer to initialize its CMAC from 3 vs. 2, the agent will initially be unable to distinguish between some states and actions because the inter-task mappings allow duplication of values. The weights corresponding to the tiles that are activated for the `Pass to second closest teammate` in the source task are copied into the weights for the tiles that are activated when evaluating the target task actions `Pass to second closest teammate` and `Pass to third closest teammate`. Thus a 4 vs. 3 keeper is initially unable to distinguish between these two actions after transfer. In other words, because the values for the weights corresponding to the two 4 vs. 3 actions are the same, $Q_{(target,initial)}$ will evaluate both actions as having the same expected return. The 4 vs. 3 agents will therefore have to learn to differentiate these two actions as they learn in the target task.

ρ_{RBF} is constructed similarly to ρ_{CMAC}. The main difference between the RBF and CMAC function approximators are how weights are summed together to produces values, but the weights have similar structure in both function approximators. For a given state variable, a CMAC sums one weight per tiling. An RBF differs in that it sums multiple weights for each tiling, where weights are multiplied by the Gaussian function $\phi(x - c_i)$. Thus ρ_{RBF} copies weights following the same schema as in ρ_{CMAC} in Algorithm 6.

5.2.2 Constructing ρ_{ANN}

This section discusses constructing a transfer functional for a neural network function approximator. Specifically, a learned ANN for 3 vs. 2 Keepaway is used to initialize a ANN for 4 vs. 3 Keepaway. However, the algorithm discussed generalizes to other tasks, provided an appropriate inter-task mapping.

To construct a (fully connected, feedforward) neural network for the 4 vs. 3 Keepaway target task, we first create a new 19-20-4 ANN. The weights connecting inputs 1–13 to the hidden nodes are copied over from the source task (13-20-3) network. Likewise, the weights from hidden nodes to outputs 1–3 are copied without modification. Weights from inputs 14–19 to the hidden nodes correspond to the new state variables and are copied over from the analogous 3 vs. 2 state variables, according to χ_X. The weights from the hidden nodes to the novel output are copied over from the analogous 3 vs. 2 action, according to χ_A. Every weight in the 19-20-4 network is thus set to an initial value based on the trained 13-20-3 network. Algorithm 7 describes this process for an arbitrary pair of source and target neural networks, where $ANN_{target}.weight(n_i, n_j)$ is the weight of a link between nodes i and j in the target task network, and $ANN_{source}.weight(n_i, n_j)$ is the weight of a link between nodes i and j in the source task network.

We define the function ψ to map target task nodes to nodes in the source task ANN:

$$\psi(n) = \begin{cases} \chi_X(n), & \text{if node } n \text{ is an input} \\ \chi_A(n), & \text{if node } n \text{ is an output} \\ \delta(n), & \text{if node } n \text{ is a hidden node} \end{cases}$$

where a function δ represents a mapping between hidden nodes. In our experiments, the number of hidden nodes used are the same in both tasks, resulting in the mapping: ψ(“n^{th} *hidden node in the target task ANN*”) = “n^{th} *hidden node in the source task ANN*.” However, if the target task ANN had additional hidden nodes, a more sophisticated mapping could be utilized, such as distributing the link weights connecting source task hidden nodes over multiple weights connecting target task hidden nodes.

Whereas ρ_{CMAC} and ρ_{RBF} copied many weights (hundreds or thousands, where increasing the amount of source task training will increase the number of learned, non-zero, weights), ρ_{ANN} always copies the same number of weights regardless of training. In fact, ρ_{ANN} initializes only 140 novel weights for this pair of network topologies in the 4 vs. 3 task (in addition to the 320 weights that existed in 3 vs. 2). Thus, this transfer functional is in some sense simpler than the other two ρs.

Algorithm 7. APPLICATION OF ρ_{ANN}

1: Create target task network, ANN_{target}, with specified topology
2: **for** each pair of nodes n_i, n_j in ANN_{target} **do**
3:     **if** $ANN_{source}.weight(\psi(n_i), \psi(n_j))$ is defined **then**
4:         $ANN_{target}.weight(n_i, n_j) \leftarrow ANN_{source}.weight(\psi(n_i), \psi(n_j))$

5.3 Empirical Evaluation of Value Function Transfer

Having introduced inter-task mappings and Value Function Transfer, this section experimentally evaluates Value Function Transfer in the Keepaway domain. These results constitute one of the core contributions of this monograph by empirically showing that this approach to TL can significantly improve learning in a complex RL domain (that was not specifically designed for transfer).

Section 5.3.1 presents experiments that demonstrate Value Function Transfer can significantly improve learning in 4 vs. 3 by transferring from 3 vs. 2, using the inter-task mappings introduced in Section 5.1.1 and the transfer functionals from Sections 5.2.1 and 5.2.2. Section 5.3.2 analyzes Value Function Transfer further by experimenting with a number of variations, such as changing how ρ_{CMAC} is constructed, how the inter-task mapping is defined, and CMAC initialization values. Section 5.3.3 then shows that Value Function Transfer can be successfully applied to players in different Keepaway tasks, even if those players have different kick actuators. Value Function Transfer is successfully used for cross-domain transfer between Knight Joust and 4 vs. 3 Keepaway in Section 5.3.4. Section 5.3.5 uses variants of 3 vs. 2 Keepaway in experiments that show how transfer efficacy is reduced when the source and target task become less similar, even to the point of negative transfer. Lastly, Section 5.3.6 presents results that show Value Function Transfer scales to larger Keepaway tasks, and investigates *multi-step transfer* [Taylor et al. (2007a)] to learn such complex tasks. The experiments in this section are summarized in Table 5.2.

5.3.1 Transfer from 3 vs. 2 Keepaway to 4 vs. 3 Keepaway

As discussed previously (Section 1.2.1), there are many possible ways to measure the benefit of transfer. In this chapter, we measure TL performance by recording how much training target task keepers require to reach a threshold performance in various situations. In order to quantify how fast an agent in 4 vs. 3 learns, we set a target performance of 10.0 seconds per episode for ANN learners, while CMAC and RBF learners have a target of 11.5 seconds. These threshold times are chosen so that learners are able to consistently attain the performance level without transfer, but players using transfer must also learn and do not initially perform above the threshold. CMAC and RBF learners are able to learn better policies than the ANN learners and thus have higher threshold values. When a group of four CMAC keepers has learned to keep the ball from the three takers for an average of 11.5 seconds

Table 5.2 This table summarizes the primary Value Function Transfer experiments in this section of the monograph. Keepaway tasks are abbreviated in this table by the number of players (i.e., "3 vs. 2" is substituted for "3 vs. 2 Keepaway"). The experiments in Section 5.3.3, marked with an $*$, focus on transfer between agents with different actuators.

Method Name: Value Function Transfer			
Scenario: Value Function Transfer is only applicable when both the source task agent and target task agent use TD learning, and both use the same type of function approximation. Results suggest that learning speeds in target tasks are improved more than with other TL methods for TD RL agents. All experiments use Sarsa as the base RL algorithm.			
Source Task	**Target Task**	**Function Approximator**	**Section**
3 vs. 2	4 vs. 3	CMAC, RBF, or ANN	5.3.1
3 vs. 2*	4 vs. 3*	CMAC	5.3.3
Knight Joust	4 vs. 3	Tabular and CMAC	5.3.4
3 vs. 2 Flat Reward	4 vs. 3	CMAC	5.3.5
3 vs. 2 Giveaway	4 vs. 3	CMAC	5.3.5
3 vs. 2	5 vs. 4	CMAC	5.3.6
3 vs. 2 – 4 vs. 3	5 vs. 4	CMAC	5.3.6
5 vs. 4	6 vs. 5	CMAC	5.3.6
6 vs. 5	7 vs. 6	RBF	5.3.6
3 vs. 2 – 6 vs. 5	7 vs. 6	RBF	5.3.6

over 1,000 episodes, we say that the keepers have successfully learned the 4 vs. 3 task. Thus agents learn until the on-line reward of the keepers, averaged over 1,000 episodes (with exploration), passes a set threshold.[2] In 4 vs. 3, it takes a set of four keepers using CMAC function approximators 30.8 simulator hours[3] (roughly 15 hours of wall-clock time, or 12,000 episodes) on average to learn to possess the ball for 11.5 seconds when training without transfer. By comparison, in 3 vs. 2, it takes a set of three keepers using CMAC function approximators 5.5 hours on average to learn to maintain control of the ball for 11.5 seconds when training without transfer.

[2] We begin each trial by following the initial policy for 1,000 episodes without learning (and therefore without counting this time towards the learning time). This enables us to assign a well-defined initial performance when we begin learning because there already exist 1,000 episodes to average over.

[3] All times reported in this section refer to simulator time, which is roughly twice that of the wall clock time. We only report simulator time (a surrogate for sample complexity) and not computational complexity for Keepaway; the running time for Sarsa and the application of ρ is negligible compared to that of the RoboCup Soccer Server.

The ANN used in 4 vs. 3 is a 19-20-4 feedforward network. The ANN learners do not learn as quickly nor achieve as high a performance before learning plateaus. After training four keepers using ANN function approximation without transfer in 4 vs. 3 for over 80 hours, the average hold time was only 10.3 seconds.

Using the three ρs described earlier, keepers in 4 vs. 3 Keepaway can initialize their action-value functions from $\rho(Q_{(3vs2,final)})$. We do not claim that these initial action-value functions are correct (and empirically they are not), but instead that the constructed action-value functions allow the learners to more quickly discover better-performing policies. This section's results compare learning 4 vs. 3 without transfer to learning 4 vs. 3 after using Value Function Transfer with varying amounts of 3 vs. 2 training. Analyses of learning times required to reach threshold performance levels show that agents utilizing CMAC, RBF, and ANN function approximation are all able to learn faster in the target task by using Value Function Transfer with ρ_{CMAC}, ρ_{RBF}, and ρ_{ANN}, respectively.[4]

To test the effect of using Value Function Transfer, we train a set of keepers for some number of 3 vs. 2 episodes on a $25m^2$ field, save the function approximator's weights ($Q_{(3vs2,final)}$) from a random 3 vs. 2 keeper[5], and use the weights to initialize all four keepers in 4 vs. 3 on a $25m^2$ field so that $Q_{(4vs3,initial)} \leftarrow \rho(Q_{(3vs2,final)})$. The agents train on the 4 vs. 3 Keepaway task until reaching the performance threshold.

To determine if keepers using CMAC function approximation can benefit from transfer we compare the time it takes agents to learn the target task after transferring from the source task with the time it takes to learn the target task without transfer. Table 5.3 reports the average time spent training in 4 vs. 3 players using CMAC function approximation to achieve an 11.5 second average possession time after different amounts of 3 vs. 2 training. The minimum learning times are in bold. To overcome the high amounts of noise in our evaluation we run at least 25 independent trials for each data point reported. The top row of the table shows the average time needed to learn the target task without transfer.

Column three reports the time spent training in the target task. The target task training time TL scenario goal is met by any TL experiments that take less time than learning without transfer. The fourth column shows the total time spent training in the source task and target task. The 5th column shows the standard deviation of the training time. As can be seen from the table, spending time training in the 3 vs. 2 task can cause the total learning time to decrease. Rows where transfer reduces the time in the third column, relative to the time needed to learn without TL, denote

[4] Our results hold for other threshold times as well, provided that the threshold is not initially reached without training and that learning will enable the keepers' performance to eventually cross the threshold.

[5] We do so under the hypothesis that the policy of a single keeper represents all of the keepers' learned knowledge. Though in theory the keepers could be learning different policies that interact well with one another, so far there is no evidence that they do. One pressure against such specialization is that the keepers' start positions are randomized. There may be such specialization when each keeper starts in the same location every episode. Informal experiments that used all source task action-value functions did not produce significant differences in performance.

Table 5.3 Results show that learning Keepaway with a CMAC and Value Function Transfer can reduce training time (in simulator hours). Minimum learning times for reaching the 11.5 second threshold are bold. As source task training time increases, the required target task training time decreases. The total training time is minimized with a moderate amount of source task training.

CMAC Learning Results

# 3 vs. 2 Episodes	Ave. 3 vs. 2 Time	Ave. 4 vs. 3 Time	Ave. Total Time	Std. Dev.
0	0	30.84	30.84	4.72
10	0.03	24.99	25.02	4.23
50	0.12	19.51	19.63	3.65
100	0.25	17.71	17.96	4.70
250	0.67	16.98	**17.65**	4.82
500	1.44	17.74	19.18	4.16
1000	2.75	16.95	19.70	5.5
3000	9.67	9.12	18.79	2.73
6000	21.65	**8.56**	30.21	2.98

experiments where Value Function Transfer successfully reduces the total training time.

The potential of Value Function Transfer is evident in Table 5.3, which is visually depicted in Figure 5.2. The table and chart both show that the time to threshold metric for the target time scenario and the total time scenario can be improved by Value Function Transfer. To analyze these results, we conduct a number of Student's t-tests to determine if the differences between the distributions of learning times for the different settings are significant.[6] These tests confirm that all the differences in the distributions of 4 vs. 3 training times when using Value Function Transfer are statistically significant ($p < 10^{-16}$), compared to training 4 vs. 3 without transfer. Not only is the time to train the 4 vs. 3 task decreased when we first train on 3 vs. 2, but the total training time is less than the time to train 4 vs. 3 without transfer. We conclude that in the Keepaway domain, training first on a simple source task can increase the rate of learning enough that the total training time is decreased when using a CMAC function approximator.

Analogous experiments for Keepaway players using RBF and ANN function approximation are presented in Table 5.4 and as bar charts in Figures 5.3 and 5.4.

[6] Throughout this monograph, Student's t-tests are used to determine statistical significance. T-tests can accurately estimate whether two distributions of samples are statistically different, even for small sample sizes, but assumes that the distributions are normally distributed. When the distributions contain a large numbers of samples (typical $n >= 30$), the distribution can be assumed to be normal by the Central Limit Theorem. However, for smaller sample sizes, we technically should test whether the distribution is normal, or use a non-parametric test, such as the Mann-Whitney U test. The U statistic was calculated to verify that differences in Table 5.3 were significant, as well as for an experiment in Section 7.1.3 with many fewer data points, and in all cases the outcome was similar to that of the Student's t-test. Thus we use the simpler (and more common) t-test throughout the remainder of this monograph to determine statistical significance.

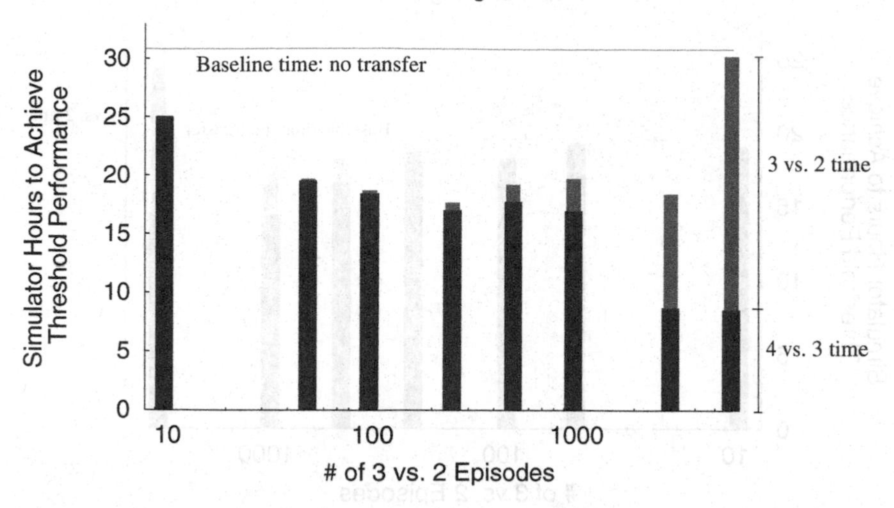

Fig. 5.2 This figure graphs the results in Table 5.3 using a logarithmic scale on the x-axis. The thin bars show the amount of time spent training in the source task, the thick bars show the amount of time spent training in the target task, and their sum represents the total time. Note that the target task time bar is graphed below the source task time bar to make the trend of decreasing target task time apparent.

Table 5.4 Results from learning Keepaway with different amounts of 3 vs. 2 training time (in simulator hours) indicate that ρ_{RBF} and ρ_{ANN} can reduce training time for RBF players (11.5 second threshold) and ANN players (10.0 second threshold). Minimum learning times for each method are in bold.

RBF and ANN Learning Results

# of 3 vs. 2 Episodes	Ave. RBF 4 vs. 3 Time	Ave. RBF Total Time	Standard Deviation	Ave. ANN 4 vs. 3 Time	Ave. ANN Total Time	Standard Deviation
0	19.52	19.52	6.03	33.08	33.08	16.14
10	18.99	19.01	6.88	19.28	**19.31**	9.37
50	19.22	19.36	5.27	22.24	22.39	11.13
100	18.00	18.27	5.59	23.73	24.04	9.47
250	18.00	18.72	7.57	22.80	23.60	12.42
500	16.56	18.12	5.94	19.12	20.73	8.81
1,000	**14.30**	**17.63**	3.34	**16.99**	20.19	9.53
3,000	14.48	26.34	5.71	17.18	27.19	10.68

Successful transfer is again demonstrated as transfer agents' target task training time and total training time are less than the time required to learn the target task without transfer. All numbers reported are averaged over at least 25 independent trials. All transfer experiments with RBF players that use at least 500 3 vs. 2 episodes show a statistically significant difference from those that learn without transfer ($p < 9.1 \times 10^{-3}$), while the learning trials that used less than 500 source task episodes

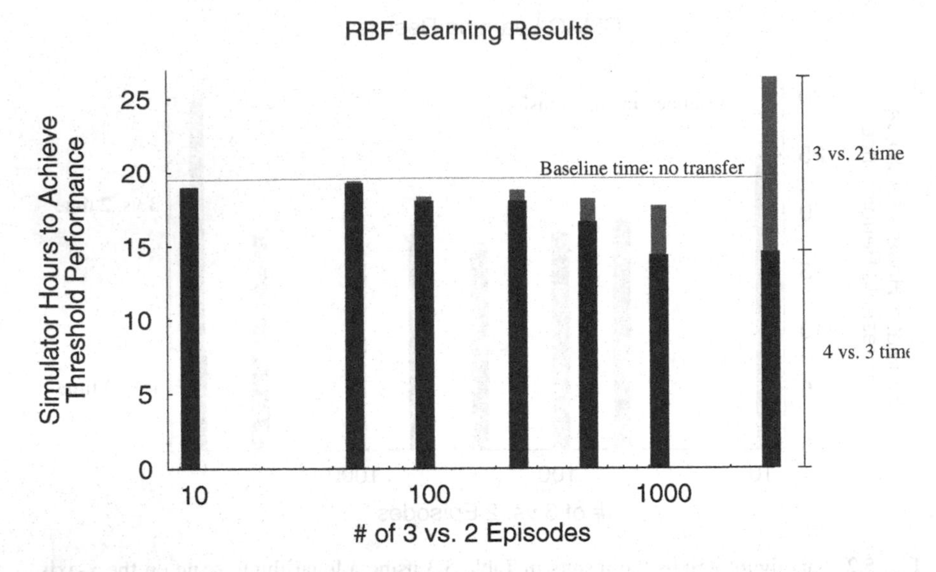

Fig. 5.3 This chart shows the RBF results from Table 5.4, where the x-axis uses a logarithmic scale, the thick bars show 4 vs. 3 learning times, the thin bars show the 3 vs. 2 time, and the combination shows the total time.

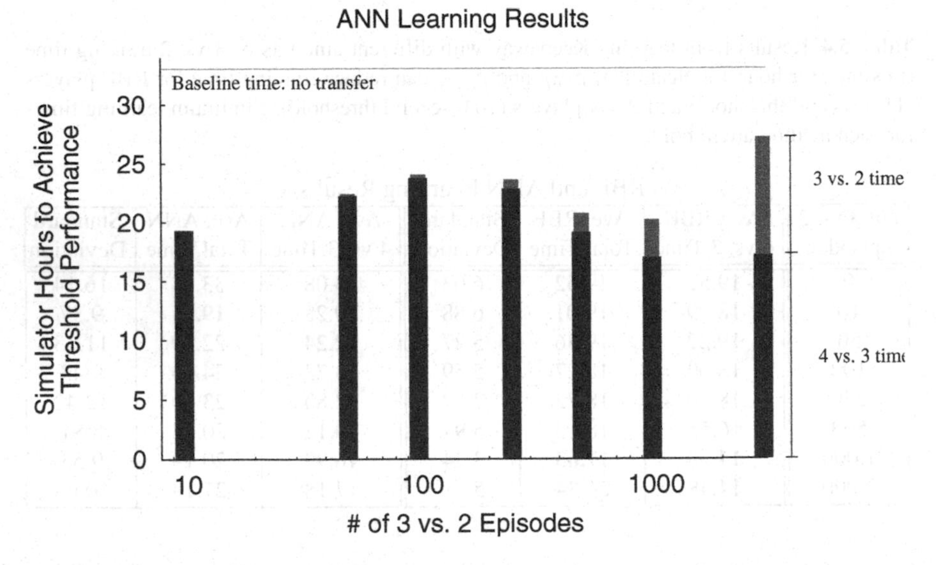

Fig. 5.4 This bar chart shows the neural network results from Table 5.4, where the x-axis uses a logarithmic scale, the thick bars show 4 vs. 3 learning times, the thin bars show the 3 vs. 2 time, and the combination shows the total time.

did not significantly reduce the target task training time. The target task training time is significantly reduced via Value Function Transfer in all ANN experiments ($p < 1.6 \times 10^{-3}$).

The RBF function approximator yielded the best learning rates for 3 vs. 2 Keepaway, followed by the CMAC function approximator, and lastly the ANN trained with backpropagation. However, Value Function Transfer provided the lowest percentage speedup to the RBF agents. One hypothesis is that transfer is less useful to the best learners. If a particular representation is poorly suited for a task, it could be that transfer is able to provide proportionally more speedup because it is that much further from an "optimal learner." Nonetheless, while some function approximators get more or less benefit from Value Function Transfer, it is clear that all three are able learn the target task faster with the technique, and that more training in the source task generally reduces the time needed to learn the target task.

5.3.2 *Understanding ρ_{CMAC}'s Benefit*

To better understand how Value Function Transfer uses ρ_{CMAC} to reduce the required training time in the target task, and to isolate the effects of its various components, this section details a number of supplemental experiments.[7]

To help understand how ρ_{CMAC} enables transfer its two components are isolated. First, the transfer functional is ablated so that the final averaging step, which places the average weight into all zero weights (Algorithm 6, lines 8–11), is removed. As anticipated, the benefit from transfer is increasingly degraded, relative to using the full ρ_{CMAC}, as fewer numbers of training episodes in the source task are used. The resulting 4 vs. 3 training times are all shorter than training without transfer, but longer than when the averaging step is incorporated. The averaging step appears to give initial values to weights in the state/action space that have never been visited when the source task agents are only allowed a limited number of learning episodes. In this scenario, the full ρ_{CMAC} can provide a useful bias in the target task even with very little 3 vs. 2 training. If agents are given more time to explore the source task, more weights in the source task CMAC are set, and the difference between the full transfer functional and the ablated transfer functional is reduced. This set of experiments, shown in Table 5.6, show that the averaging step is most useful with small amounts of source task training, but becomes less so as more source experience is accumulated.

In order to evaluate a ρ_{CMAC} that *only* performed the averaging step, we first learn for a number of episodes in 4 vs. 3, instead of 3 vs. 2, and then overwrite all zero weights with the non-zero average. Thus, 4 vs. 3 is both the source and target task;

[7] Informal experiments showed that the CMAC and RBF transfer results were qualitatively similar, which is reasonable given the two function approximator's many similarities. Thus we expect that the supplemental experiments in this section would yield qualitatively similar results if we used RBFs rather than CMACs. While results demonstrate that all three function approximators can successfully transfer knowledge, these supplementary experiments focus on CMAC function approximation so that this work is more comparable to previous work in Keepaway [Stone et al. (2005)] that also used CMACs.

Table 5.5 This table lists results that show transfer using the full ρ_{CMAC}, using ρ_{CMAC} without the final averaging step (rows 4 and 5), using only the averaging step of ρ_{CMAC} (rows 6 and 7), and when averaging weights in the source task before transferring the weights (row 8). Note that all variants of ρ_{CMAC} are able to successfully reduce the required 4 vs. 3 training time relative to learning without transfer.

Ablation Studies with ρ_{CMAC}

Transfer Functional	# Source Task Episodes	Ave. Target Task Time	Standard Deviation
No Transfer	0	30.84	4.72
ρ_{CMAC}	100	17.71	4.70
ρ_{CMAC}	1000	16.95	5.5
ρ_{CMAC}	3000	9.12	2.73
$\rho_{CMAC,No\ Averaging}$	100	25.68	4.21
$\rho_{CMAC,No\ Averaging}$	3000	9.53	2.28
only averaging	100	19.06	6.85
only averaging	3000	10.26	2.42
$\rho_{CMAC,Ave\ Source}$	1000	15.67	4.31

Table 5.6 10 independent trials are averaged for different values for initial CMAC weights. The top line is reproduced from Table 5.3. None of the trials with initial weights of 1.0 were able to reach the 11.5 threshold within 45 hours, and thus are shown as N/A above.

Time Required for CMAC 4 vs. 3 players to Achieve 11.5 sec. Performance

Initial CMAC weight	Ave. Learning Time	Standard Deviation
0	30.84	4.72
0.5	35.03	8.68
1.0	N/A	N/A
Each weight randomly selected from a uniform distribution from [0,1.0]	28.01	6.93

our results report the time spent learning the 4 vs. 3 task after transfers. Isolating the averaging step helps determine how important this step is to ρ_{CMAC}'s effectiveness. Applying the averaging step causes the total training time to decrease below that of training 4 vs. 3 without transfer, but again the training times are longer than using the full ρ_{CMAC} on weights trained in 3 vs. 2. This result confirms that both parts of ρ_{CMAC} contribute to reducing 4 vs. 3 training time and that training on 3 vs. 2 is more beneficial for reducing the required 4 vs. 3 training time than training on 4 vs. 3 and applying ρ_{CMAC} (see Table 5.5, rows 6 and 7, for result details).

The averaging step in ρ_{CMAC} is defined so that the average weight in the *target* CMAC overwrites all zero-weights. We also conducted a set of 30 trials which modified ρ_{CMAC} so that the average weight in the *source* CMAC is put into all zero-weights in the target CMAC, which is possible when agents in the source task know that their saved weights will be used for Value Function Transfer. Table 5.5 shows that when the weights are averaged in the source task ($\rho_{CMAC,Ave\ Source}$, bottom row) the performance is not statistically different ($p > 0.05$) from when using ρ_{CMAC} for Value Function Transfer with the same number of source task episodes.

To verify that the 4 vs. 3 CMAC players benefit from Value Function Transfer and not from simply having non-zero initial weights, we initialized CMAC weights uniformly to 0.5 in one set of experiments, to 1.0 in a second set of experiments, and then to random numbers uniformly distributed from 0.0-1.0 in a third set of experiments. We do so under the assumption that 0.0, 0.5, and 1.0 are all reasonable initial values for weights (although in practice 0.0 seems most common). The learning time was never statistically better than learning with weights initialized to zero, and in some experiments the non-zero initial weights decreased the speed of learning. This set of experiments suggests that haphazardly initializing CMAC weights may hurt the learner, but systematically setting them through Value Function Transfer may be beneficial. We conclude that the benefit of transfer is not a byproduct of our initial setting of weights in the CMAC (see Table 5.6 for result details).

To test the sensitivity of the ρ_{CMAC} function, we next change it in two different ways. We first define $\rho_{modified}$ by modifying χ_A so that instead of mapping the novel target task action `Pass to third closest keeper` into the action `Pass to second closest keeper`, we map the novel action to the source task action `Hold ball`. Now $Q_{4vs3,initial}$ will initially evaluate `Pass to third closest keeper` and `Hold ball` as equivalent for all states. We also define $\rho_{partial}$ by defining new inter-task mappings. χ_A and χ_X are modified so that distances, angles, and actions not present in 3 vs. 2 are not initialized in the target task. Using these partial inter-task mappings, $\rho_{partial}$ is a functional which copies over information learned in 3 vs. 2 exactly but assigns the average weight to all novel state variables and actions in 4 vs. 3.

When $\rho_{modified}$ was used by Value Function Transfer to initialize weights in 4 vs. 3, the total training time increases relative to the normal ρ_{CMAC}, but still outperforms training without transfer. Similarly, $\rho_{partial}$ is able to outperform learning without transfer, but underperforms the full ρ_{CMAC}, particularly for higher amounts of training in the source task. Choosing non-optimal inter-task mappings when constructing ρ seems to have a detrimental, but not necessarily disastrous, effect on the training time. These different transfer functionals significantly outperform learning without transfer. Results in Table 5.7 show that the structure of ρ is indeed important to the success of transfer.

When the CMACs' weights are loaded into the keepers in 4 vs. 3, the initial hold times of the keepers do not differ significantly from those of keepers with uninitialized CMACs (i.e., CMACs where all weights are initially set to zero). The information contained in the function approximators' weights prime the 4 vs. 3 keepers to more quickly learn their task by biasing their search, even though the knowledge transferred is of limited initial value. See Figures 5.5 and 5.6 for representative learning curves and Table 5.8 for details. The more similar the source and target tasks are, the more of a jumpstart in performance would be expected. For example, in the degenerate case where the source and target task are identical, the initial performance in the target task will be equivalent to the final performance in the source

Table 5.7 Results show that transfer with the full ρ_{CMAC} outperforms using suboptimal or partial inter-task mappings. The top line represents learning without transfer and lines 1–3 are from Table 5.3.

Testing Sub-optimal Inter-task Mappings

Transfer Functional	# of 3 vs. 2 Episodes	Ave. 4 vs. 3 Time	Standard Deviation
No Transfer	0	30.84	4.72
ρ_{CMAC}	100	17.71	4.70
ρ_{CMAC}	3000	9.12	2.73
$\rho_{modified}$	100	21.74	6.91
$\rho_{modified}$	3000	10.33	3.21
$\rho_{partial}$	100	18.90	3.73
$\rho_{partial}$	3000	12.00	5.38

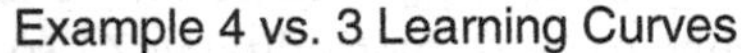

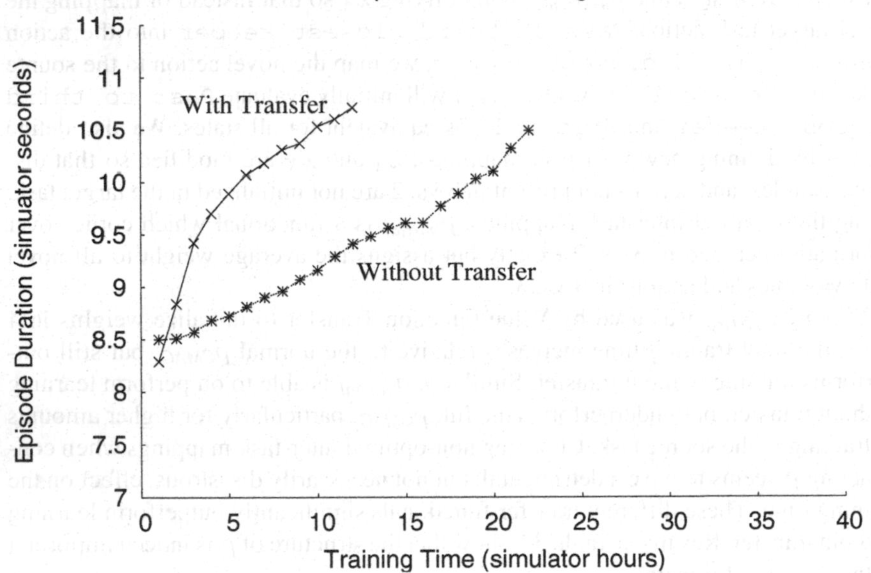

Fig. 5.5 The average performance of representative learning curves show that transfer via inter-task mapping does not produce a significant jumpstart in 4 vs. 3, but enables faster learning. Learning in 4 vs. 3 without transfer is compared with learning after transfer from 250 episodes of 3 vs. 2.

task. However, in such a situation, reducing the total time — the more difficult transfer scenario — would prove impossible.

Value Function Transfer relies on effectively reusing learned data in the target task. We hypothesized that successfully leveraging this data may be affected by ε,

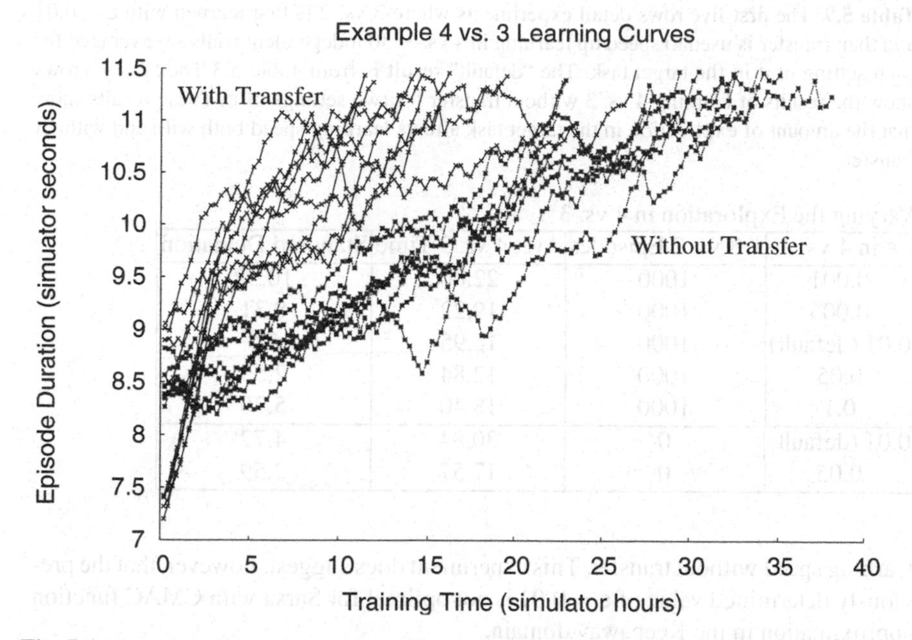

Fig. 5.6 Eight 4 vs. 3 learning curves without transfer are compared to eight learning curves in 4 vs. 3 after transferring from 250 episodes of 3 vs. 2. This graph shows the high variance in trials graphed in Figure 5.5.

Table 5.8 This table shows the difference in initial performance between 4 vs. 3 players with and without transfer. 40 independent trials are averaged for each setting and the differences in initial performance (i.e., initial episode lengths) are small. A Student's t-test shows that 8.46 and 8.92 are not statistically different ($p > 0.05$) while 8.46 and 9.24 are ($p < 6.6 \times 10^{-5}$).

Initial Performance in 4 vs. 3 with CMAC Function Approximation

# of 3 vs. 2 Episodes	Ave. Performance (sec.)	Standard Deviation
0	8.46	0.17
1000	8.92	1.48
6000	9.24	1.15

the ϵ-greedy exploration parameter, which balances exploration with exploitation. Recall that we had chosen an exploration rate of 0.01 (1%) in Keepaway to be consistent with past research. Table 5.9 lists the results of learning 3 vs. 2 Keepaway with $\epsilon = 0.01$ for 1,000 episodes, utilizing ρ_{CMAC}, and then learning 4 vs. 3 Keepaway with various settings for ϵ. The results show that of these 4 additional settings for ϵ, only $\epsilon = 0.05$ is statistically better than the default rate of 0.01. To further explore this last result we ran a series of 30 trials that learned 4 vs. 3 without transfer with the value of $\epsilon = 0.05$ and found that there was a significant difference from learning 4 vs. 3 without transfer with $\epsilon = 0.01$. The speedup for this particular setting of ϵ in transfer, relative to the default value, is explained by the increased

Table 5.9 The first five rows detail experiments where 3 vs. 2 is first learned with $\varepsilon = 0.01$ and then transfer is used to speed up learning in 4 vs. 3. 30 independent trials are averaged for each setting of ε in the target task. The "default" result is from Table 5.3 The last two rows show the results of learning 4 vs. 3 without transfer for two settings of ε. These results show that the amount of exploration in the target task affects learning speed both with and without transfer.

Varying the Exploration in 4 vs. 3

ε in 4 vs. 3	# 3 vs. 2 Episodes	Ave. 4 vs. 3 Time	Standard Deviation
0.001	1000	22.06	10.52
0.005	1000	19.22	8.31
0.01 (default)	1000	16.95	5.5
0.05	1000	12.84	2.55
0.1	1000	18.40	5.70
0.01 (default)	0	30.84	4.72
0.05	0	17.57	2.59

learning speed without transfer. This experiment does suggest, however, that the previously determined value of $\varepsilon = 0.01$ is not optimal for Sarsa with CMAC function approximation in the Keepaway domain.

From the experiments in this section we conclude:

1. Both parts of ρ_{CMAC} — copying weights based on χ_X and χ_A, and the final averaging step — contribute to the success of Value Function Transfer. The former gives more benefit as more training is completed in the source task and the second helps more when less knowledge is gained in the source task before transfer.
2. Using Value Function Transfer with ρ_{CMAC} is superior to weights initialized to zero (training without transfer), as well as weights initialized to 0.5, 1.0 and [0,1.0], three other reasonable initial settings.
3. A suboptimal or incomplete transfer functional, such as $\rho_{modified}$ and $\rho_{partial}$, allows Value Function Transfer to improve learning speeds relative to learning without transfer, but not as much as the full ρ_{CMAC}.
4. Keepers initialized by Value Function Transfer in the source task do not always initially outperform players learning without transfer in the target task, but the transfer players are able to learn faster.

5.3.3 *Transfer between Players with Differing Abilities*

Results in the previous sections demonstrate that Q-values can be successfully transferred between 3 vs. 2 Keepaway and 4 vs. 3 Keepaway. This section tests how robust Value Function Transfer is to changes in agents' abilities. In addition to changing the number of players between the source and target tasks, other variables such as the size of the field, the wind speed, and the players' abilities can be modified in the Keepaway domain. It is a qualitatively different challenge to use Value Function Transfer to speed up learning between two tasks where the agents' actions have qualitatively different effects (i.e., the nature of the transition function, T, has been modified) in addition to different state and action spaces. We choose to test the robustness of Value Function Transfer by changing the passing actuators on some sets of agents so that the passes are less accurate.[8] We show in this section that Value Function Transfer speeds up learning, relative to learning without transfer, in the following three scenarios:

1. Learning 4 vs. 3 with **inaccurate** passing actuators after transferring from 3 vs. 2 keepers with **inaccurate** passing actuators.
2. Learning 4 vs. 3 with a **normal** passing actuators after transferring from 3 vs. 2 keepers with **inaccurate** passing actuators.
3. Learning 4 vs. 3 with **inaccurate** passing actuators after transferring from 3 vs. 2 keepers with **normal** passing actuators.

Recall that accurate CMAC players learning without transfer in 4 vs. 3 take 30.1 hours to reach the threshold performance level (row 1 of Table 5.10). When sets of CMAC keepers learn 4 vs. 3 without transfer while using the less accurate pass mechanism, the average time to reach an average performance of 11.5 seconds is 54.2 hours (row 4 of Table 5.10). ρ_{CMAC} can speed up learning in the target task when both the target and source tasks have inaccurate actuators (rows 5 and 6 of Table 5.10, scenario #1). These two results, as well as all other 4 vs. 3 transfer learning times in this table, are statistically significant when compared to learning the relevant 4 vs. 3 task without transfer ($p < 1.4 \times 10^{-12}$).

Now consider scenario #2. Suppose that the 4 vs. 3 target task players use inaccurate passing and some 3 vs. 2 keepers have trained using an accurate pass action in the source task. As we can see in the third group of results in Table 5.10 (rows 7 and 8), even though the players in the source task have different actuators than in the target task, transfer is able to significantly speed up learning compared to not using transfer.

This result confirms that the same ρ will allow Value Function Transfer to effectively transfer between tasks that not only have different S and A, but also have qualitatively different actions. This situation is of practical import, as many robotic systems experience gradual degradation in performance over time due to wear and tear. If a set of robots with worn down actuators are available, they may still be able to benefit from action-value function transfer of Q-values from learners that have fresh actuators. Alternately, if a set of agents have learned a task and then later want to learn a new target task but have damaged their actuators since learning the source task, transfer may still increase the speed of learning.

In the complementary experiment (scenario #3), agents in the source task have inaccurate actuators and agents in the target task have normal actuators. We perform Value Function Transfer after 500 and 3,000 episodes of 3 vs. 2 with inaccurate passing to initialize the Q-values of agents in 4 vs. 3 with accurate passing. The final two rows in Table 5.10 again show that using transfer is a significant improvement over

[8] Actuators are changed in the benchmark players by changing the pass action from the default "PassNormal" to "PassFast." This increases the speed of the passed ball by 50% and significantly reduces pass accuracy.

Table 5.10 Results showing transfer via inter-task mapping benefits CMAC players utilizing ρ_{CMAC} with two types of actuators. These results demonstrate that transfer can succeed even when actions in the source and target tasks are qualitatively different. The results in rows 1–3 are from Table 5.3.

Learning Results with Different Actuators

# of 3 vs. 2 Episodes	3 vs. 2 Time	3 vs. 2 Actuator Accurate?	Ave. 4 vs. 3 Time	Standard Deviation	4 vs. 3 Actuator Accurate?
0	0	N/A	30.84	16.14	Yes
500	1.44	Yes	17.74	4.16	Yes
3000	9.67	Yes	9.12	2.73	Yes
0	0	N/A	54.15	6.13	No
500	1.23	No	37.3	9.24	No
3000	8.36	No	29.86	9.20	No
500	1.37	Yes	37.54	7.48	No
3000	9.45	Yes	24.17	5.54	No
500	1.3	No	18.46	3.93	Yes
3000	8.21	No	13.57	3.64	Yes

learning without transfer. This results suggests that a fielded agent with worn down actuators would be able to successfully transfer its learned action-value function to agents with undamaged actuators. Interestingly, transferring from source task keepers that have accurate actuators is more effective than transferring from source task keepers that have inaccurate actuators, regardless of which actuator is used in the target task. This is most likely because it is easier to learn with accurate actuators, which means that there is more useful information to transfer from agents trained with accurate actuators.

Section 5.3.1 first showed that transfer from 3 vs. 2 keepers with accurate pass actuators to 4 vs. 3 keepers with accurate pass actuators was successful. This section demonstrates that transfer also works in three different scenarios that consider source and/or target agents with inaccurate passing actuators. Combined, these results show that Value Function Transfer is able to speed up learning in multiple target tasks with different state and action spaces, and even when the agents have qualitatively different actuators (and therefore have qualitatively different transition functions) in the two tasks.

5.3.4 *Transfer from Knight Joust to 4 vs. 3 Keepaway*

Previous sections in this chapter have empirically demonstrated that Value Function Transfer can successfully transfer between different Keepaway tasks. A more difficult challenge is to transfer between different domains. Such *cross-domain transfer* [Taylor and Stone (2007a)] has been a long-term goal of transfer learning because it could allow transfer between less similar tasks, significantly increasing TL's flexibility. In previous sections, experiments showed that Value Function Transfer

Table 5.11 This table describes the mapping between state variables and actions from 4 vs. 3 to Knight Joust. Note that the we have made Jump West in the Knight Joust correspond to passing to K_2 and Jump East correspond to passing to K_3, but these options are symmetrically equivalent, as long as the state variables and actions are consistent.

χ_X: 4 vs. 3 to Knight Joust

4 vs. 3 state variable	Knight Joust state variable
$dist(K_1, T_1)$	$dist(P, O)$
$\text{Min}(ang(K_2, K_1, T_1), ang(K_2, K_1, T_2), ang(K_2, K_1, T_3))$	$ang(West)$
$\text{Min}(ang(K_3, K_1, T_1), ang(K_3, K_1, T_2), ang(K_3, K_1, T_3))$	$ang(East)$
$\text{Min}(ang(K_4, K_1, T_1), ang(K_4, K_1, T_2), ang(K_4, K_1, T_3))$	$ang(East)$
All other Keepaway variables	ø

χ_A: 4 vs. 3 to Knight Joust

4 vs. 3 action	Knight Joust action
`Hold ball`	`Forward`
`Pass to closest teammate`	`Jump`$_{West}$
`Pass to second closest teammate`	`Jump`$_{East}$
`Pass to third closest teammate`	`Jump`$_{East}$

can reduce the total training time, in part because source tasks were selected so that they were faster to learn than the target tasks. If a source task is selected from a different domain, we may be able to select source tasks that take orders of magnitude less training time, potentially reducing the total training time even more significantly.

The Knight Joust task, introduced in Section 4.5, is less similar to 4 vs. 3 Keepaway than 3 vs. 2 Keepaway is. There are many fewer state variables, a less similar transition function, a fully observable state, the learner's actions are deterministic, and the reward structure is very different. Experiments in this section show that Value Function Transfer can successfully transfer between the Knight Joust task in the gridworld domain and 4 vs. 3 Keepaway in the RoboCup Soccer domain. Table 5.11 describes the inter-task mappings used to transfer between Knight Joust and 4 vs. 3 Keepaway. Our hypothesis was that the Knight Joust player would learn Q-values to represent behaviors like "move North when possible," and "jump to the side when necessary," which could be similar to holding the ball in Keepaway when possible and passing when necessary.

We use a variant of ρ_{CMAC}, which we designate $\rho_{tabular,CMAC}$, to transfer the learned weights because Knight Joust is learned with a tabular function approximator rather than a CMAC. This representation choice results in changes to the syntax of the transfer functional, described in Algorithm 8. Note that this new variant of the transfer functional is not necessitated by the novel source task. If Knight Joust were learned with a CMAC, the original ρ_{CMAC} would be sufficient for transfer between Knight Joust and 4 vs. 3.

The results in Table 5.12 report the average results from experiments with 30 independent trials each. The time spent learning 4 vs. 3 after transfer from 25,000

Algorithm 8. $\rho_{tabular,CMAC}$

1: $n_{source} \leftarrow$ number of variables in source task
2: **for** each non-zero Q-value, q_i, in the source task's Q-table **do**
3: $a_{source} \leftarrow$ action corresponding to q_i
4: **for** each state variable, x_{source}, in source task **do**
5: **for** each value x_{target} such that $\chi_X(x_{target}) = x_{source}$ **do**
6: **for** each value a_{target} such that $\chi_A(a_{target}) = a_{source}$ **do**
7: $j \leftarrow$ the tile in the target CMAC activated by x_{target}, a_{target}
8: $w_j \leftarrow (q_i/n_{source})$
9: $w_{Average} \leftarrow$ average value of all non-zero weights in the target CMAC
10: **for** each weight w_j in the target CMAC **do**
11: **if** $w_j = 0$ **then**
12: $w_j \leftarrow w_{Average}$

Table 5.12 Results from using Knight Joust to speed up learning in 4 vs. 3 Keepaway. Knight Joust is learned with Q-learning and tabular function approximation and Keepaway players are learned using Sarsa with CMAC function approximation. The results in Row 1 are from Table 5.3. Both transfer times are significantly less than learning without transfer, as determined via Student's t-tests ($p < 0.05$).

Transfer from Knight Joust into 4 vs. 3

# of Knight Joust Episodes	Ave. 4 vs. 3 Time	Standard Deviation
0	30.84	4.72
25,000	24.24	16.18
50,000	18.90	13.20

or 50,000 episodes of Knight Joust are statistically different from learning without transfer (determined via Student's t-tests). Recall that the wall-clock time of the Knight Joust simulator is negligible when compared to the either wall-clock or simulator time and thus, in practice, 4 vs. 3 time is roughly the same as total time.

The reader may notice that the number of source task episodes used in these experiments is much larger than other experiments in this chapter. The reason for this is two-fold. First, Knight Joust is learned with tabular function approximation, which is significantly slower to learn than a CMAC because there is no generalization. Second, because the wall-clock time requirements for this domain were so small, we felt justified in allowing the source task learners run until learning plateaued (which takes roughly 50,000 episodes).

The main importance of these results is that they show that Value Function Transfer can succeed between tasks with very different dynamics. Keepaway has stochastic actions, is partially observable, and uses a continuous state space. In contrast, Knight Joust has no stochasticity in the player's actions, is fully observable, and has a discrete state space.

Table 5.13 Results compare transferring from three different source tasks. Each row is an average of 30 independent trials. The 3 vs. 2 Flat Reward task improves performance relative to learning without transfer, but less than when transferring from Keepaway. The 3 vs. 2 Giveaway task can decrease 4 vs. 3 performance when it is used as a source task. Rows 1–3 are replicated from Table 5.3.

Transfer into 4 vs. 3 with ρ_{CMAC}: Different Source Tasks

Source Task	# of 3 vs. 2 Episodes	Ave. 4 vs. 3 Time	Ave. Total Time	Std. Dev.
none	0	30.84	30.84	4.72
Keepaway	1000	16.95	19.70	5.5
Keepaway	3000	9.12	18.79	2.73
Flat Reward	1000	25.11	27.62	6.31
Flat Reward	3000	19.42	28.03	8.62
Giveaway	1000	27.05	28.58	10.71
Giveaway	3000	32.94	37.10	8.96

5.3.5 Variants on 3 vs. 2 Keepaway: Negative Transfer

In this section we introduce two novel variants of the 3 vs. 2 Keepaway task to show how Value Function Transfer with ρ_{CMAC} can *fail* to improve performance relative to learning without transfer. Results in this section provide a cautionary tale: if source and target tasks are not very similar, transfer can fail to decrease, or can even increase, learning times.

3 vs. 2 is first modified so that the reward is defined as +1 for each action, rather than +1 for each timestep. Players in the novel *3 vs. 2 Flat Reward* [Taylor et al. (2007a)] task can still learn to increase the average episode time. We hypothesized that the changes in reward structure would hinder Value Function Transfer into 4 vs. 3 because the different reward structures would not only produce a different optimal policy, but the returns predicted by the action-value function would be less similar to correct predictions in 4 vs. 3 Keepaway.

3 vs. 2 is next modified so that the reward is defined to be -1 for each timestep. The novel task of *3 vs. 2 Giveaway* [Taylor et al. (2007a)] is, in some sense, the opposite of Keepaway. Given the available actions, the optimal action is always for the player closest to the takers to hold the ball until the takers captures it. We hypothesized that using Value Function Transfer from Giveaway to 4 vs. 3 Keepaway would produce *negative transfer*, where the required target task training time is increased by using transfer.

Table 5.13 shows the results of using these two 3 vs. 2 variants as source tasks. The table also compares these transfer results to using the standard 3 vs. 2 task for transfer, and learning without transfer. Transfer from the Flat Reward tasks gives a benefit relative to learning without transfer, but not nearly as much as transferring from 3 vs. 2 Keepaway. Student's t-tests determine that transfer after 3,000 episodes of Giveaway is significantly slower than learning without transfer.

5.3.6 Larger Keepaway Tasks and Multi-Step Transfer

Sections 5.3.1–5.3.5 considered experiments that use variants of 3 vs. 2 as a source task (with the notable exception of Knight Joust) and variants of 4 vs. 3 as a target task. This section explores how well Value Function Transfer scales by testing it on increasingly complex tasks. As discussed in Chapter 1, one of the potential benefits of transfer is to make learning on large problems tractable. With this goal in mind, experiments in this section consider 5 vs. 4, 6 vs. 5 and 7 vs. 6 Keepaway. In addition to showing that one-step transfer is possible, as was done in previous sections, we also investigate multi-step transfer and show that in some cases it is superior to one-step transfer.

5 vs. 4 Keepaway is harder than 4 vs. 3 Keepaway, as discussed in Section 4.3.1. In 5 vs. 4, a set of five keepers using CMAC function approximators require an average of 59.9 hours (roughly 24,000 episodes) to learn to maintain possession of the ball for 11.5 seconds when training without transfer. We first show that our TL method can be used to speed up the 5 vs. 4 Keepaway task, which provides evidence for scalability to larger tasks. In addition to using 4 vs. 3 to speed up learning in 5 vs. 4, we show that Value Function Transfer can be applied *twice* to learn the 3 vs. 2, 4 vs. 3, and 5 vs. 4 tasks in succession.

Results in Table 5.14 demonstrate that Value Function Transfer scales to the 5 vs. 4 Keepaway task. The 5 vs. 4 task has been successfully learned when the 5 keepers are able to possess the ball for an average of 9.0 seconds over 1,000 episodes. ρ_{CMAC} can be formulated by extending χ_X and χ_A so that they can transfer the action-value function from 4 vs. 3 to 5 vs. 4, analogous to the way it transfers values from 3 vs. 2 to 4 vs. 3. These results are shown in rows 2 and 3 of Table 5.14.

Recall that χ_X and χ_A can also be formulated so that Value Function Transfer can transfer from 3 vs. 2 to 5 vs. 4 (Section 5.1.1). Table 5.14, rows 4 and 5, show that this mapping formulation is successful. In fact, there is more benefit than transferring from 4 vs. 3. We posit that this is because it is easier to learn more in the simpler source task, outweighing the fact that 5 vs. 4 is less related to 3 vs. 2 than it is to 4 vs. 3. Another way to understand this result is that in a fixed amount of experience, players in 3 vs. 2 are able to update more weights than 4 vs. 3, measured as a percentage of the total possible number of weights used in the task.

We may also use multi-step transfer, a two-step application of Value Function Transfer, to learn 5 vs. 4. The learned action-value function from 3 vs. 2 is used to initialize the action-value function in 4 vs. 3 after applying ρ_{CMAC}. After training, the final 4 vs. 3 action-value function is then used to initialize the action-value function for 5 vs. 4. Using this procedure we find that the time to learn 5 vs. 4 is reduced to roughly 27% of learning without transfer (Table 5.14, bottom row). A t-test confirms that the differences between all 5 vs. 4 training times shown are statistically significant ($p < 3.6 \times 10^{-7}$) when compared to learning without transfer.

These results show that Value Function Transfer allows 5 vs. 4 Keepaway to be learned faster after training on 4 vs. 3 and/or 3 vs. 2. They also suggest that a multi-step process, where tasks are made incrementally more challenging, may produce faster learning times than a single application of Value Function Transfer. A similar

Table 5.14 Results showing that learning Keepaway with a CMAC and applying Value Function Transfer can reduce training time (in simulator hours) for CMAC players in 5 vs. 4 with a target performance of 9.0 seconds. All numbers are averaged over at least 25 independent trials.

CMAC Learning Results in 5 vs. 4 Keepaway

# of 3 vs. 2 Episodes	# of 4 vs. 3 Episodes	Ave. 5 vs. 4 Time	Ave. Total Time	Standard Deviation
0	0	22.58	22.58	3.46
0	500	13.44	14.60	7.82
0	1000	9.66	12.02	4.50
500	0	6.76	**8.18**	1.90
1000	0	6.70	9.66	2.12
500	500	**6.19**	8.86	1.26

Table 5.15 10 independent trials are averaged for learning 6 vs. 5 with and without transfer from 5 vs. 4. The threshold performance time is 8.0 seconds. A Student's t-test confirms that the difference is statistically significant ($p < 2.2 \times 10^{-11}$).

Transfer in 6 vs. 5 with CMAC Function Approximation

# of 5 vs. 4 Episodes	Ave. 6 vs. 5 Time	Ave Total Time	Standard Deviation
0	22.85	22.85	1.71
1000	9.38	11.53	2.38

χ_X, χ_A, and ρ_{CMAC} can be constructed to significantly speed up learning in 6 vs. 5 as well (also taking place on a $25m \times 25m$ field), as shown in Table 5.15.

Figure 5.7 shows that Value Function Transfer further scales to the more difficult task of 7 vs. 6 Keepaway. This experiment uses the latest version of the RoboCup Soccer Server (version 11.1.2) and RBF function approximation, and thus is not directly comparable to the previous CMAC experiments. Agents learning without transfer are able to improve their performance, but learning is much slower than in simpler Keepaway tasks. The "Transfer from 6 vs. 5" trails show that learning 6 vs. 5 for 4,000 episodes and then using Value Function Transfer can successfully speed up learning in 7 vs. 6. In fact, the initial performance in 7 vs. 6 immediately after transfer in not attained by the non-transfer learners within 70 simulator hours. (The trails transferring from 6 vs. 5 begin with a performance of 7.7 ± 0.2 seconds per episode at 8.7 simulator hours, whereas the trials that do not use transfer only attain a performance of 7.2 ± 0.4 seconds per episode after 70.0 simulator hours.) The "Multistep Transfer" learning curve shows the results of learning 3 vs. 2, 4 vs. 3, 5 vs. 4, and 6 vs. 5 for 1,000 episodes each (using Value Function Transfer between each sequential pair of tasks) and then transferring into 7 vs. 6. Unlike the experiment in 5 vs. 4, this experiment does not show a significant improvement by multi-step transfer, but multi-step transfer again shows a significant improvement over no transfer.

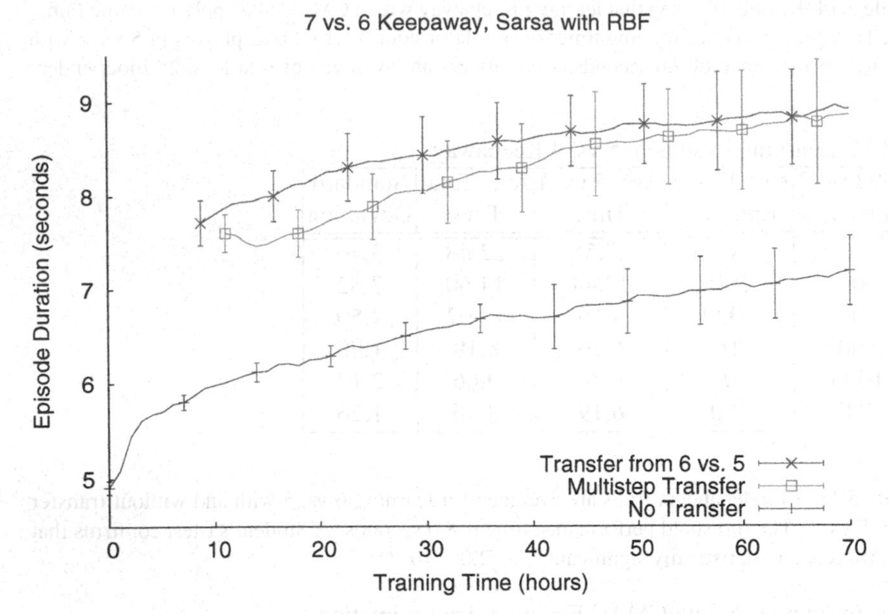

Fig. 5.7 Learning 7 vs. 6 Keepaway with transfer from 4,000 episodes of 6 vs. 5 is superior to learning without transfer. Multi-step transfer (1000 episodes each of 3 vs. 2, 4 vs. 3, 5 vs. 4, and 6 vs. 5) outperforms learning without transfer, but is not as effective as 1-step transfer. This graph shows the total time, causing the transfer learning curves to be shifted to the right by the total amount of training time spent in the source task(s). Error bars show the standard deviation and each learning curve averages 15 independent learning trials.

5.4 On the Applicability of Value Function Transfer

The methods and results in this chapter constitute the core contribution of this monograph. In particular, the results presented in this chapter serve as an existence proof that Value Function Transfer *can* be effectively used with inter-task mappings to speed up learning in a target task after training in a source task.

It is important to recognize the domain knowledge contained in χ_X and χ_A that assists in generating an effective ρ. As experiments show, simply copying weights without respecting the inter-task mapping is not a viable method of transfer, as our function approximator representations necessarily differ between the two tasks due to changes in S and A. Simply putting the average value of the 3 vs. 2 weights into the 4 vs. 3 function approximator does not give nearly as much benefit as using a ρ which explicitly handles the different state and action values. Likewise, when we used $\rho_{modified}$ (introduced in Section 5.3.2), which copied the values for the weights corresponding to the some of the state variables incorrectly, learning in 4 vs. 3 was significantly slower. These results suggest that a ρ which is able to leverage inter-task similarities will outperform more simple ρs.

At this point, the reader may reasonably ask, "even if a correct inter-task mapping is provided, in what situations is Value Function Transfer guaranteed to work?" As discussed in Chapter 3, no TL methods for RL tasks currently provide guarantees for transfer efficacy in the general case, although we discuss some possibilities for such proofs in the future work section of this monograph (Section 9.3). Instead, the current state of the art relies on heuristics to decide when transfer may be effective and which types of source tasks may be effectively used for a given target task. When considering a scenario that has a human in the loop, such assumptions are reasonable; in our experience human intuition has been an effective guide for TL methods. Selecting a source task or determining an inter-task mapping may be thought of as another way to bias the agent, such as when a researcher selects state variables, a function approximator, a particular learning method, or a shaping reward.

It is still important to attempt to qualify where a given TL method will and will not be useful. Value Function Transfer relies on an inter-task mapping between similar states and actions. In order to be effective, the mapping should identify state variables and actions that have similar effects on the long-term discounted reward. If, for instance, the Keepaway task were changed so that instead of receiving a reward of $+1$ at every time step, the agent received a $+10$, the ρ could be trivially modified so that all the weights were multiplied by 10. However, if the reward structure is more significantly changed to that of *Giveaway*, ρ would need to be dramatically changed, if it could be effectively formulated at all.[9]

We hypothesize that the main requirement for successful Value Function Transfer is that at least one of the following is true, on average:

1. The actions with the highest return in $Q_{(source,final)}(\chi_X(s_{target}),\cdot)$ are among the best actions in the target task's optimal policy at that state: $\pi^\star(s_{target})$.
2. The average Q-values learned in the source task's action-value function are of the same magnitude as Q-values in the trained target task's action-value function.

The first condition will work to bias the learner so that the best actions in the target task are chosen more often, even if these actions' Q-values are incorrect. The second condition will make learning faster because smaller adjustments to the function approximators' weights will be needed to reach their optimal values, relative to not using transfer, even if the optimal actions are not initially chosen.[10] In this chapter, an example of the first condition being met is that a keeper learns to hold the ball in the source task until forced to pass. `Hold ball` is often the correct action in both 3 vs. 2 Keepaway and 4 vs. 3 Keepaway when the takers are far away from the ball.

[9] The "obvious" solution of multiplying all weights by -1 would not work. In Keepaway a keeper typically learns to hold the ball until a taker comes within roughly $6m$. Thus, if all weights from this policy were multiplied by -1, the keepers would continually pass the ball until a taker came within $6m$. These Giveaway episodes would last much longer than simply forcing the first keeper to the ball to always hold, which is very easily learned.

[10] This condition is most important for incremental learners, but does not necessarily apply to batch learners. For instance, suppose that Q-values transferred to the target task allowed the agent to follow the optimal policy because the relative values of actions were correct (a best-case scenario for the first condition). When the agent acts in the target task, it will collect samples following this optimal policy. Then, when it updates its action-value function, it can compute a more accurate Q-value function, regardless of the magnitude of the transferred Q-values.

The second condition is also met between 3 vs. 2 and 4 vs. 3 by virtue of similar reward structures and roughly similar episode lengths. If either of these conditions were not true, the transfer functional we employed would have to account for the differences (or suffer from reduced transfer efficacy). In Section 5.3.5, our transfer experiment from the 3 vs. 2 Flat Reward task to the 4 vs. 3 Keepaway task showed reduced transfer efficacy (compared to transfer from the standard 3 vs. 2 Keepaway task) because the first condition did not always hold and the second condition was violated. Transfer from 3 vs. 2 Giveaway into 4 vs. 3 Keepaway showed negative transfer because both of the above assumptions were violated.

This chapter has also set up a number of questions that will be addressed in the following chapters. Specifically:

1. Can other types of knowledge be transferred between learners effectively, other than an action-value function? (Yes, as discussed in Chapters 6 and 7.)
2. If an agent has a selection of source tasks, can it determine which task is more likely to be useful for transfer into a given target? (Yes, but there are no guarantees, as discussed in Section 7.1.)
3. Can inter-task mappings be learned, rather than hand-coded? (Yes, but they are not always as effective as hand-coded mappings — see Chapter 8.)

Chapter 6
Extending Transfer via Inter-Task Mappings

The previous chapter introduced inter-task mappings and empirically demonstrated that Value Function Transfer could use such mappings to significantly improve the speed of reinforcement learning. This chapter presents two additional transfer methods that conceptually build upon the Value Function Transfer method and also utilize inter-task mappings.

Section 6.1 introduces *Q-Value Reuse* [Taylor et al. (2007a)], which allows agents to directly reuse a learned action-value from a source task in a target task. The primary benefit of this method is that it allows TD learners in the source task and target task to used different function approximators by saving the source task agent's final action-value function and leveraging it, without modification, in the target task.

Policy Transfer [Taylor et al. (2007b)] is discussed in Section 6.2. Both Value Function Transfer and Q-Value Reuse require that agents in the source task and target task utilize action-value function learning methods. In some tasks, direct policy search methods may outperform TD methods (c.f. [Taylor et al. (2006)]). Policy Transfer is a method that can effectively transfer neural network action selectors (i.e., policies) between a source task and a target task. Again, inter-task mappings can be used to enable transfer if the tasks have different state variables and/or actions.

6.1 Q-Value Reuse

The three transfer functionals introduced for value-function transfer (see Sections 5.2.1 and 5.2.2) are specific to particular function approximators. In this section we introduce a more flexible approach, *Q-Value Reuse*, to transfer between tasks. Rather than initializing a function approximator in the target task with values learned in the source task, we reuse the entire learned source task's action-value function. One potential benefit is increased flexibility: an agent in the target task may train using a function approximator different from that used by the source task agent.[1] A second benefit is that no explicit transfer step to copy weights is needed.

[1] If the task is discrete, the agent may also use a tabular representation, which can be treated as a type of function approximator.

M.E. Taylor: Transfer in Reinforcement Learning Domains, SCI 216, pp. 121–138.
springerlink.com

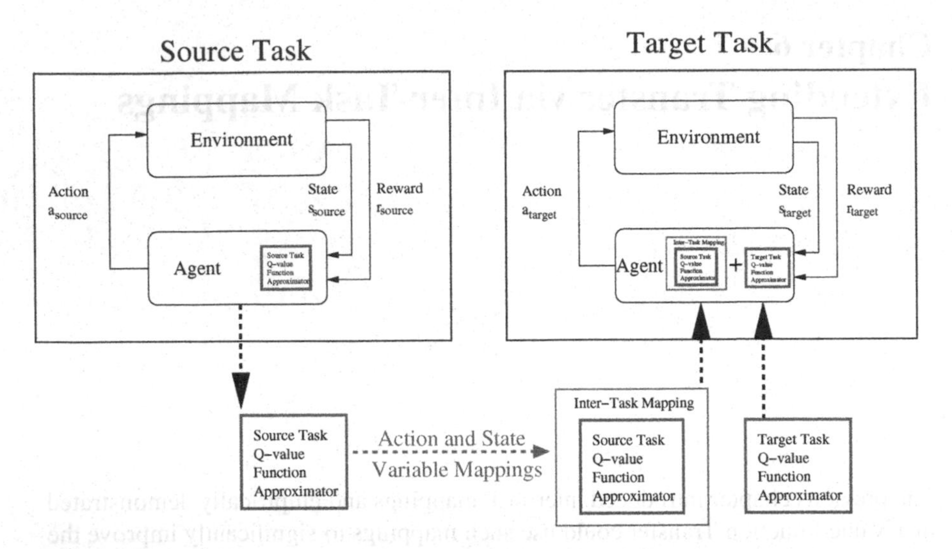

Fig. 6.1 In Q-Value Reuse, the source task function approximator is saved after being learned in the source task. An agent in the target task then uses the saved function approximator, in conjunction with inter-task mappings, to evaluate states and actions. To learn, a target task agent modifies a separate target task Q-value function approximator, leaving the transferred source task function approximator unchanged.

As we will see later (Section 6.1.1), this is important when the target task function approximator has a much larger number of weights than the source task function approximator.

The insight behind Q-Value Reuse is that a copy of the source task's function approximator can be retained and used to calculate the source task's Q-values for state, action pairs ($Q_{source} : S_{source} \times A_{source} \mapsto \mathbb{R}$) (see Figure 6.1). When computing Q-values for the target task, the target task agent first maps the state and actions via inter-task mappings. The computed Q-value in the target task agent is a combination of the output of the source task's saved (and unchanging) function approximator applied to the transformed (s,a), and the target task's modifiable function approximator applied to the target task states and actions:

$$Q(s,a) = Q_{sourceFA}(\chi_X(s), \chi_A(a)) + Q_{learningFA}(s,a) \tag{6.1}$$

Sarsa updates in the target task are computed as normal, but only the target function approximator's weights ($Q_{learningFA}$) are eligible for updates. Q-Value Reuse is similar to *reward shaping* [Colombetti and Dorigo (1993), Mataric (1994)] in that we directly use the predicted rewards from the source task to bias the learner in the target task. Note that if $\chi_X(s)$ or $\chi_A(a)$ were undefined for an (s,a) pair in the target task, $Q(s,a)$ would simply equal $Q_{learningFA}(s,a)$.

Potential drawbacks of this transfer method include an increased lookup time and larger memory requirements, relative to Value Function Transfer. Such requirements will grow linearly in the number of transfer steps; while they are not substantial with

Table 6.1 This table summarizes the primary Q-Value Reuse experiments in this section of the monograph

Method Name: Q-Value Reuse			
Scenario: Applicable when the source and target task agents use TD learning. Agents are not required to use the same function approximator. This method is not as effective as Value Function Transfer and may require extra memory due to multiple function approximators, but it is more flexible as agents may use different function approximators. All experiments use Sarsa learning.			
Source Task	**Target Task**	**Function Approximator**	**Section**
Standard 2D Mountain Car	Standard 3D Mountain Car	CMAC	6.1.1
3 vs. 2 Keepaway	4 vs. 3 Keepaway	CMAC	6.1.2

a single source task, they may become prohibitive when using multiple source tasks or when performing multi-step transfer (e.g., Section 5.3.6). Table 6.1 summarizes the Q-Value Reuse experiments presented in the next two sections.

6.1.1 Q-Value Reuse Results: Mountain Car

After learning the Standard 2D Mountain Car task with a CMAC (as discussed in Section 4.1.1) for 100 episodes, there are an average of 2,400 weights set in the CMAC. If we then used Value Function Transfer to initialize a learner in the Standard 3D Mountain Car task, there would a total of 459,300 weights because all tiles in the 4D CMAC would be initialized.[2] If a CMAC trains in the Standard 3D Mountain Car task for 10,000 episodes, only 208,200 weights are used, on average. One of the motivations of using Q-Value Reuse is to avoid such an overhead due to initializing weights which will likely not be used by the target task agent.

When Q-Value Reuse is used to transfer between the Standard 2D Mountain Car task and the Standard 3D Mountain Car task, the agent first saves its 2D CMAC after training on the source task with Sarsa. Second, in the target task, the agent modifies the weights of a 4D CMAC when learning. When computing the action-value for an (s_{target}, a_{target}) pair, the agent also uses the saved 2D CMAC to evaluate the current position. Conceptually, $Q(s_{target}, a_{target}) = Q_{2DCMAC}(\chi_X(s_{target}), \chi_A(a_{target})) + Q_{4DCMAC}(s_{target}, a_{target})$. The inter-task mappings for 3D to 2D Mountain Car are defined in Table 6.2. Actions that accelerate the car towards the goal are mapped together, while actions that accelerate the car away from the goal are mapped together. State variables are mapped from the target to the source so that they are consistent

[2] There are 9^4 weights (i.e., tiles) per 4D tiling. There are 14 tilings and 5 actions, leading to 459,270 weights that would be initialized ($2394 \times 9^4 \times 14 \times 5$). Our Sarsa implementation of Mountain Car from the RL-Glue Task Library is currently not able to handle this number of weights, making Q-Value Reuse preferable to Value Function Transfer.

Table 6.2 This table describes the mapping used transfer between the Standard 2D and Standard 3D Mountain Car tasks

3D to 2D Mountain Car Inter-Task Mappings	
Action Mapping	State Variable Mapping
χ_A(Neutral) = Neutral	$\chi_X(x) = x$
χ_A(North) = Right	$\chi_X(\dot{x}) = \dot{x}$
χ_A(East) = Right	or
χ_A(South) = Left	$\chi_X(y) = x$
χ_A(West) = Left	$\chi_X(\dot{y}) = \dot{x}$

with the action mapping (e.g., minimizing x in both the target task and the source task moves the agent closer to the goal).

Notice that the state variable inter-task mapping is many-to-one. Because of this, when calculating the contribution from the source task CMAC, the target task state must be used *twice*[3]:

$$\begin{aligned} Q(s_{target}, a_{target}) = & \; Q_{2dCMAC}(\chi_X(x_{target}), \chi_X(\dot{x}_{target}), \chi_A(a_{target})) + \\ & \; Q_{2dCMAC}(\chi_X(y_{target}), \chi_X(\dot{y}_{target}), \chi_A(a_{target})) + \\ & \; Q_{4dCMAC}(x_{target}, y_{target}, \dot{x}_{target}, \dot{y}_{target}, a_{target}) \end{aligned}$$

While learning the 3D task, the target task agent's CMAC weights are modified by Sarsa(λ) and will allow for an accurate approximation of the action-value function, even though the transferred source CMAC will not necessarily produce a significant jumpstart in the target task.

Figure 6.2 shows learning curves in 3D Mountain Car, each averaged over 25 independent trials. After each episode we evaluate the policy off-line without exploration. To graph the learning curve we average all 25 learning curves with a 10 episode window.

The line labeled "Transfer: Hand-Coded Mapping" shows the performance of agents in the 3D task that transfer an action-value function learned during 100 episodes of training on the 2D task. Student's t-tests show that transfer using hand-coded inter-task mappings (χ_X and χ_A) significantly outperform learners that do not use transfer. The "Transfer: Averaged Mapping" shows the performance of agents that again transfer an action-value function learned during 100 source task episodes, but use a different inter-task mapping. Instead of using χ_X and χ_A as defined in Table 6.2, these learners call into the saved Q_{2dCMAC} multiple times and average over *all* possible mappings. This is equivalent to a mapping that assigns a uniform prior from all actions in the target task to all actions in the source task (and likewise for state variables). The poor performance of these learners emphasizes that

[3] This also has the effect of initially increasing the expected return for every (s,a) pair in the target task. While this is appropriate for 3D Mountain Car (because the target task is significantly harder than the source and the car needs many more steps to reach the goal), in other domains it may be preferable to re-scale the output from Q_{source} based on the number of times it is called for a single (s,a) target task pair.

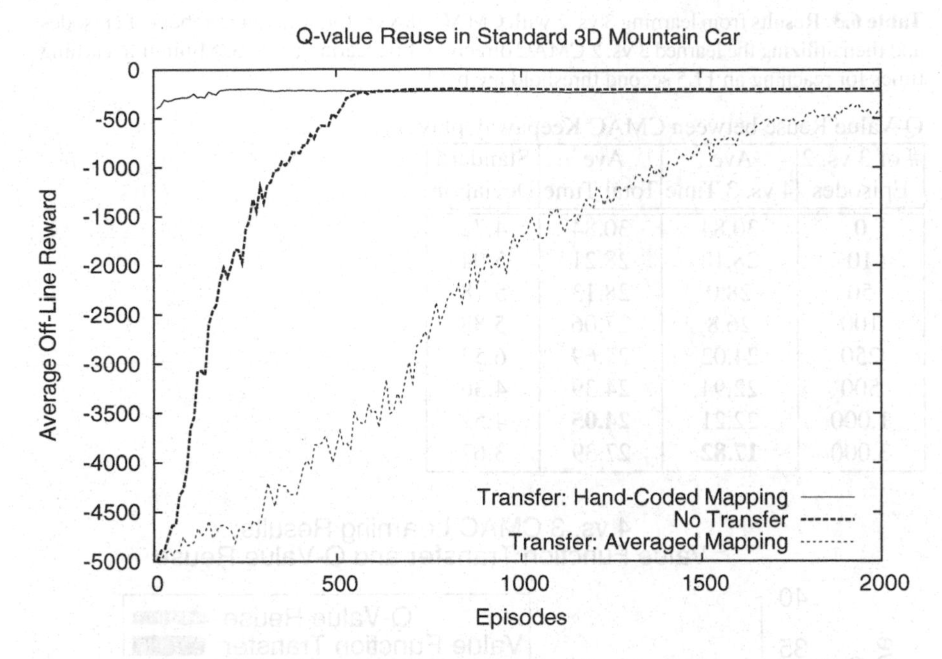

Fig. 6.2 This graph shows learning curves for learning without transfer, learning with transfer using hand-coded mappings, and learning with transfer using an averaged mapping. Each learning curve averages 25 independent trials.

haphazardly transferring without reasonable inter-task mappings can lead to negative transfer.

6.1.2 Q-Value Reuse Results: Keepaway

In this section we use Value Function Transfer with Q-Value Reuse to transfer between CMAC players in 3 vs. 2 Keepaway and 4 vs. 3 Keepaway. We utilize the same inter-task mappings as in Value Function Transfer (Section 5.1.1). When calculating $Q(s_{target}, a_{target})$, the source task CMAC effectively ignores state variables that pertain to the novel players (the 4th keeper and 3rd taker), and will evaluate the actions $\texttt{Pass}_2$ and $\texttt{Pass}_3$ as the same value. Similar to Value Function Transfer, we do not expect the initial performance in the target task to be near-optimal because of these inaccuracies. However, we do expect that learning the novel target task CMAC will be fast, relative to learning without transfer.

Table 6.3 shows the results of using Q-Value Reuse in Keepaway. Each transfer experiment shows the average of 30 independent trials. Both the target task times and the total times are statistically different from learning without transfer ($p < 0.05$, via Student's t-tests). As when using ρ_{CMAC} for Value Function Transfer (Table 5.3), spending more time learning 3 vs. 2 correlates with a decrease in the time required

Table 6.3 Results from learning 3 vs. 2 with CMAC players for different numbers of episodes and then utilizing the learned 3 vs. 2 CMAC directly while learning 4 vs. 3. Minimum learning times for reaching an 11.5 second threshold are bold.

Q-Value Reuse between CMAC Keepaway players

# of 3 vs. 2 Episodes	Ave. 4 vs. 3 Time	Ave. Total Time	Standard Deviation
0	30.84	30.84	4.72
10	28.18	28.21	5.04
50	28.0	28.13	5.18
100	26.8	27.06	5.88
250	24.02	24.69	6.53
500	22.94	24.39	4.36
1,000	22.21	**24.05**	4.52
3,000	**17.82**	27.39	3.67

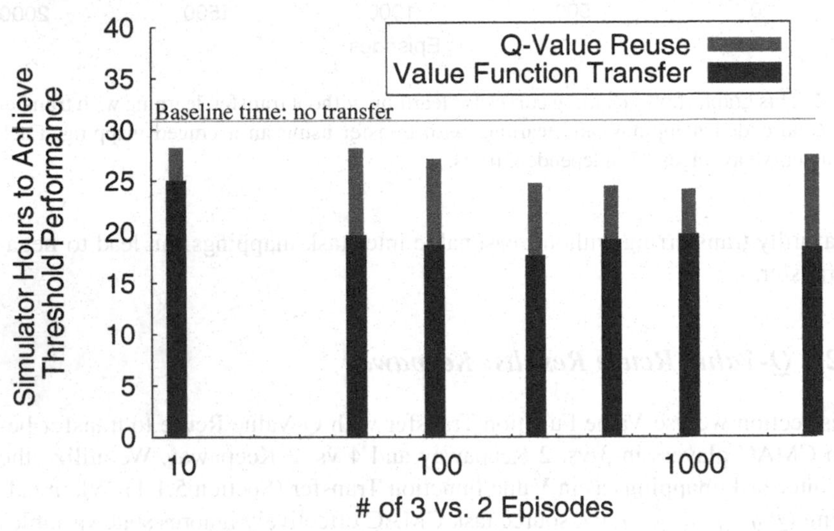

Fig. 6.3 This chart compares Value Function Transfer with Q-Value Reuse when using CMAC function approximation. The x-axis shows the number of source task episodes and the y-axis shows the resulting time the learners needed to train in the target task before reaching the threshold performance. In every set of experiments, Value Function Transfer requires less target task training time to reach the threshold performance than Q-Value Reuse does. Value Function Transfer times are repeated from Table 5.3 in Section 5.3.1 and Q-Value Reuse results are from Table 6.3.

for 4 vs. 3 players to reach an 11.5 second threshold performance. Q-Value Reuse is not as effective as Value Function Transfer (see Figure 6.3) due to the averaging step in ρ_{CMAC}. As we showed previously (Chapter 5.3.2), this averaging step has an impact on the target task learning times. However, in Q-Value Reuse we treat the source task function approximator as a "black box" and thus do not permute its values, nor use it to set the initial values of the target task's function approximator. These results suggest that if the source and target function approximators are different, or if value-function transfer would require the copying of an infeasible number of weights, Q-Value Reuse may be appropriate. However, if memory is limited, running time is critical, and/or multiple transfer steps are involved, then using Value Function Transfer may be preferable.

6.2 Policy Transfer

This monograph has only considered transfer between temporal difference RL agents up to this point. Since policy search methods, which directly search the space of policies without learning value functions, can outperform TD methods on some tasks (c.f., [Stanley and Miikkulainen (2002)] and [Taylor et al. (2006)]) extending transfer learning to policy search methods is an important goal. *Policy Transfer* is one such TL method: it transfers policies, represented as neural network action selectors, from a source task to a target task by leveraging inter-task mappings.

In this section, we first describe Policy Transfer and then empirically evaluate it in Server Job Scheduling (SJS) (described in Section 4.2) and Keepaway. Results show successful reduction in both target task training time and total training time, relative to learning without transfer.

Policy Transfer enables policies represented as neural network action selectors to be transferred between tasks (see Figure 6.4). We choose neural network action selectors because of their past successes in policy search (e.g., [Stanley and Miikkulainen (2002)], [Taylor et al. (2006)], [Whiteson and Stone (2006)]). Although we restrict the experiments to neural network action selectors (which, the reader may recall from Section 2.2.3, are direct functions from states to actions) in this monograph, there are no apparent obstacles to applying Policy Transfer to other policy search learners.

We now discuss how to construct a functional, ρ_π, for Policy Transfer such that the initial policy (or policies) in the target task can be initialized by policies learned in the source task. To perform transfer with a neural network action selector, we must convert networks trained in the source task into networks suitable for training in the target task. We cannot simply copy the policy description unaltered because, in the general case, the state and action spaces may differ between tasks, and therefore the policy function's inputs and outputs may differ.

To construct ρ_π, we use an algorithm very similar to ρ_{ANN} (introduced in Section 5.2.2). Rather than transferring a single neural network that represents a value function, ρ_π transfers one or more neural network action selectors. We again assume that two inter-task mappings are provided, χ_X and χ_A. Given χ_X, χ_A, and a trained network π_{source}, our goal is to create a new network π_{target} that can function in the

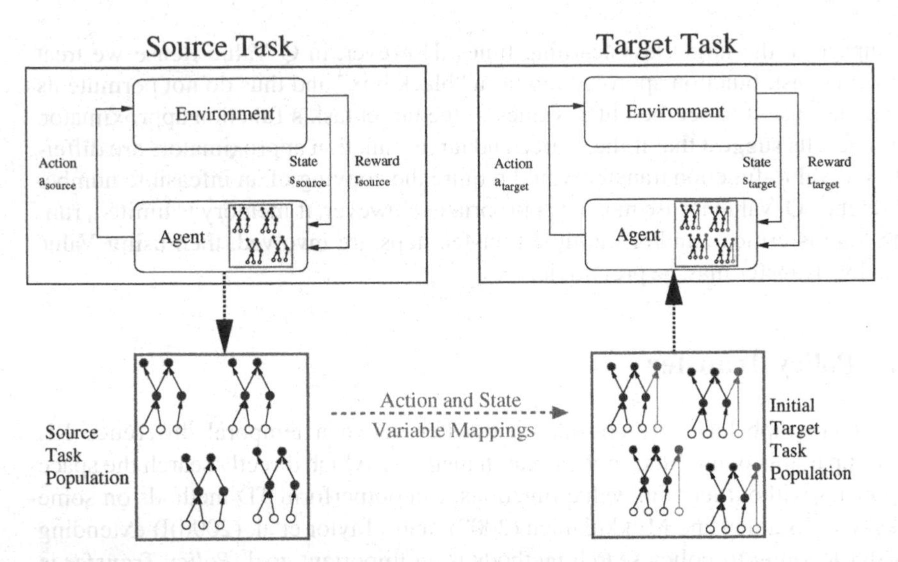

Fig. 6.4 In this monograph we use Policy Transfer in conjunction with NEAT. After learning a population of policies in the source task, the inter-task mappings are used to create new policies in the target task. These new policies, appropriate for the target task actions and state variables, are used to initialize the target task learning algorithm.

target task. Initially, we define π_{target} as a neural network with no links, one input node for each state variable in the target task, one output node for each action in the target task, and the same number of hidden nodes as in π_{source}.

We use the function ψ from Section 5.2.2 to map target task nodes to nodes in the source task action selector:

$$\psi(n) = \begin{cases} \chi_X(n), & \text{if node } n \text{ is an input} \\ \chi_A(n), & \text{if node } n \text{ is an output} \\ \delta(n), & \text{if node } n \text{ is a hidden node} \end{cases}$$

where a function δ again represents a mapping between hidden nodes. Using ψ, we can now generate π_{target} by copying the links that connect the corresponding nodes in π_{source}. For every pair of nodes n_i, n_j in π_{target}, if a link exists between $\psi(n_i)$ and $\psi(n_j)$ in π_{source}, a new link with the same weight is created between n_i and n_j.[4] By applying ρ_π to source task policies, we can initialize target task policies. All target task policies thus have structure and weights learned from the source task and we expect this knowledge to bias policies so that policy search methods can master the target task more quickly. Algorithm 9 summarizes this domain-independent process and Policy Transfer experiments are summarized in Table 6.4.

[4] Alternatively, link weights could be set such that the target network's activation for every output a, given $s_1 \ldots s_k$, is the same as the source network's activation for $\chi_A(a)$, given $\chi_X(s_1) \ldots \chi_X(s_k)$. However, informal results suggest this approach is less effective than directly copying weights.

Algorithm 9. ρ_π: Policy Transfer with NEAT

1: **for** each network π_{source} in source task population **do**
2: Construct a network π_{target} where # of input and output nodes are determined by the target task
3: Add the same number of hidden nodes to π_{target} as π_{source}
4: **for** each pair of nodes n_i, n_j in π_{target} **do**
5: **if** link$(\psi(n_i), \psi(n_j))$ in π_{source} exists **then**
6: Add link(n_i, n_j) to π_{target} with weight identical to link$(\psi(n_i), \psi(n_j))$

Table 6.4 This table summarizes the primary Policy Transfer experiments in this section of the monograph. Experiments are first conducted with full inter-task mappings and then repeated in Section 6.2.3 with partial inter-task mappings.

Method Name: Policy Transfer			
Scenario: Policy Transfer is applicable when the source and target task agents use direct policy search with ANN action selectors. All experiments use NEAT.			
Source Task	**Target Task**	**Function Approximator**	**Section**
2-job-type SJS	4-job-type SJS	ANN	6.2.1, 6.2.3
3 vs. 2 Keepaway	4 vs. 3 Keepaway	ANN	6.2.2, 6.2.3

6.2.1 *Server Job Scheduling Results*

In this section we demonstrate that Policy Transfer can successfully transfer between 2-job-type and 4-job-type scheduling (discussed in Section 4.2). Server Job Scheduling is quickly mastered with policy search learning, but is difficult for TD methods.[5] Thus Value Function Transfer is not applicable for SJS, but Policy Transfer is.

To define the inter-task mappings used for SJS (detailed in Table 6.5) we utilize a similar methodology as for Keepaway: we map target task job types 1 and 3 to source job type 1, and map target task job types 2 and 4 to source job type 2, exploiting similarities in utility curves (see Figure 6.5, a reproduction of Figure 4.7 from Section 4.2).

[5] After testing a number of different Sarsa implementations, we hypothesize that TD methods have difficultly learning an action-value function for SJS. Considering each state variable independently does not provide enough information to the learner to accurately predict the return for the 16 different actions: although the agent shows initial learning progress, performance plateaus far below that of NEAT. If the state variables are considered conjunctively, the state space becomes quite large and the agent fails to learn at all (the 16 state variables which have integral values of [0,50]). A second difficulty is that the return directly depends on the current timestep in the episode. Explicitly adding time as a state variable to the learner's state representation failed to increase performance, likely due to the added challenge of temporal generalization.

Table 6.5 This table describes the state variable mapping between the 4-job-type Server Job Scheduling task and the 2-job-type Server Job Scheduling task. State variables count the number of jobs of a particular type in different time ranges. $Count_{2,1}$ represents the number of jobs of type two that have an age of 1-50 time-steps. The action mapping is analogous to the state variable mapping. For instance, the target task action "Process the oldest job of type four in the first bin" is mapped to the source task action "Process the oldest job of type two in the first bin" (i.e., $\texttt{Process}_{4,1}$ is mapped to $\texttt{Process}_{2,1}$, just as $Count_{4,1}$ is mapped to $Count_{2,1}$).

χ_X Mapping 4-job-type SJS to 2-job-type SJS

4-job-type state variable	2-job-type state variable
$Count_{1,1}$	$Count_{1,1}$
$Count_{1,2}$	$Count_{1,2}$
$Count_{1,3}$	$Count_{1,3}$
$Count_{1,4}$	$Count_{1,4}$
$Count_{2,1}$	$Count_{2,1}$
$Count_{2,2}$	$Count_{2,2}$
$Count_{2,3}$	$Count_{2,3}$
$Count_{2,4}$	$Count_{2,4}$
$Count_{3,1}$	$Count_{1,1}$
$Count_{3,2}$	$Count_{1,2}$
$Count_{3,3}$	$Count_{1,3}$
$Count_{3,4}$	$Count_{1,4}$
$Count_{4,1}$	$Count_{2,1}$
$Count_{4,2}$	$Count_{2,2}$
$Count_{4,3}$	$Count_{2,3}$
$Count_{4,4}$	$Count_{2,4}$

After training a population of policies in 2-job-type SJS, we transfer the entire population[6] into 4-job-type SJS via Policy Transfer using ρ_π. These initial policies in the 4-job-type population will not be optimal, but should enable NEAT to more rapidly discover good policies in the target task than when learning without transfer.

During initial experiments we found that a small modification to Algorithm 9 improves transfer performance on SJS. The number of links in the target task network is double that of the source network due to different numbers of input and output nodes in the two networks. When copying weights from the source task network, we divided all values in half, significantly improving the initial performance in the target task, as well as the time to threshold, relative to using the source task weights without modification.

To test Policy Transfer in Server Job Scheduling, we first learn 2-job-type (source) task for 5 or 10 generations and save the learned population of policies

[6] Transferring the entire trained population instead of a single policy allows search to begin in the target task from a variety of locations in policy space, increasing the chances of finding a good starting point for learning. Informal results suggested that this approach was more beneficial than transferring the champion policy, or a few of the best policies.

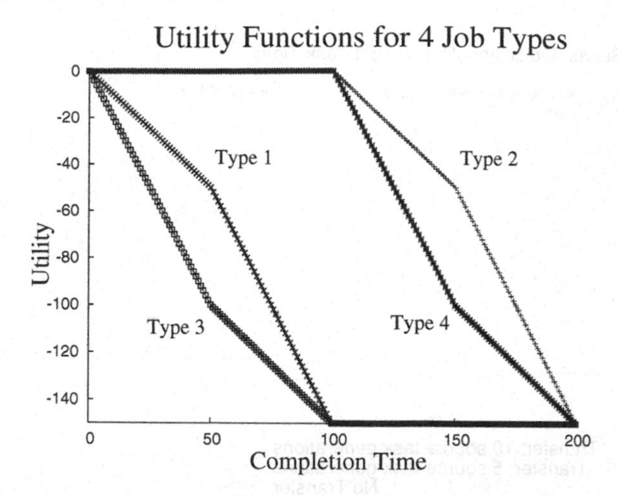

Fig. 6.5 This graph shows the four utility functions used in SJS experiments

at the end of training. We then learn the 4-job-type (target) task without transfer, and with Policy Transfer from each set of populations. Figure 6.6 shows the performance of the three sets of learning experiments vs. the number of target task generations. Transfer from 5 generations outperforms learning without transfer until generation 29 and transfer from 10 generations outperforms learning without transfer until generation 33, as determined by Student's t-tests ($p < 0.05$). Transfer from 10 generations initially performs slightly better than transfer from 5 generations, but the majority of differences are not statistically significant. Figure 6.7 shows the same data as Figure 6.6, but shows the total number of training generations. When accounting for source task training, transferring from 5 source task generations is better than transferring from 10. Transfer from 5 source task generations is statistically different from the no-transfer learning curve until generation 22 ($p < 0.05$) when accounting for the total learning time.

To better analyze the effect of Policy Transfer, we also consider a *range* of threshold values rather than a single value. By graphing multiple thresholds we can visualize how much experience a learner needs to reach a number of preset target levels and compare transfer to non-transfer learners. Additionally, performing this analysis supports our claim (in Section 1.2.1) that the particular target task performance threshold do not greatly affect the evaluation of different transfer algorithms.

Figure 6.8 shows such a graph for target task training time in SJS. The x-axis shows threshold performance values, and the y-axis shows how many target task generations were needed to achieve that performance. Each curve averages the performance of 100 independent trials and continue until one or more of the trials fails to reach the target performance (e.g., 1 of the 100 non-transfer learning curves failed to ever achieve a performance of -8,000 and the non-transfer learning curve is graphed on the range [-8750,-8050]). The no transfer learning curve is much higher than the two transfer learning curves, which equates to requiring many more

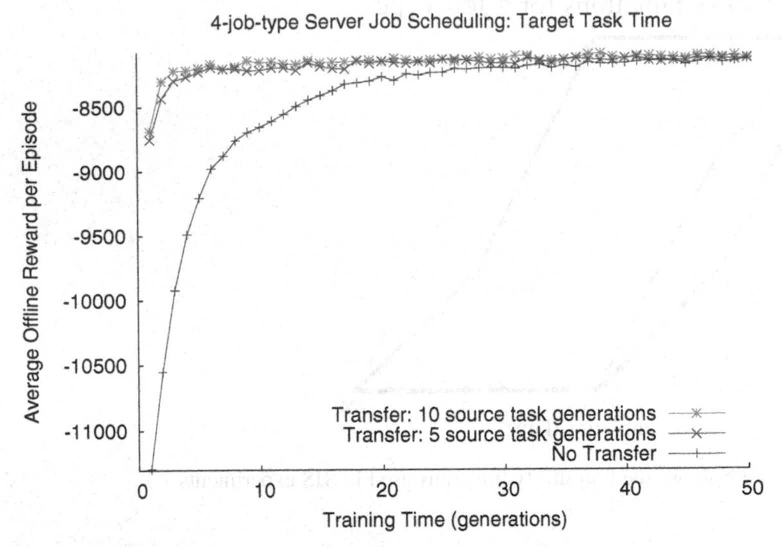

Fig. 6.6 This graph compares the performance on the 4-job-type Server Job Scheduling task without transfer, with Policy Transfer from 5 generations of 2-job-type SJS, and with Policy Transfer from 10 generations of 2-job-type SJS. The x-axis only accounts for target task training time (the source task training time is treated as a sunk cost). Each curve averages 100 independent learning trials. Recall (Section 4.2.1) that in each generation when learning SJS, the agent evaluates 50 policies for 5 episodes, resulting in 250 episodes per generation.

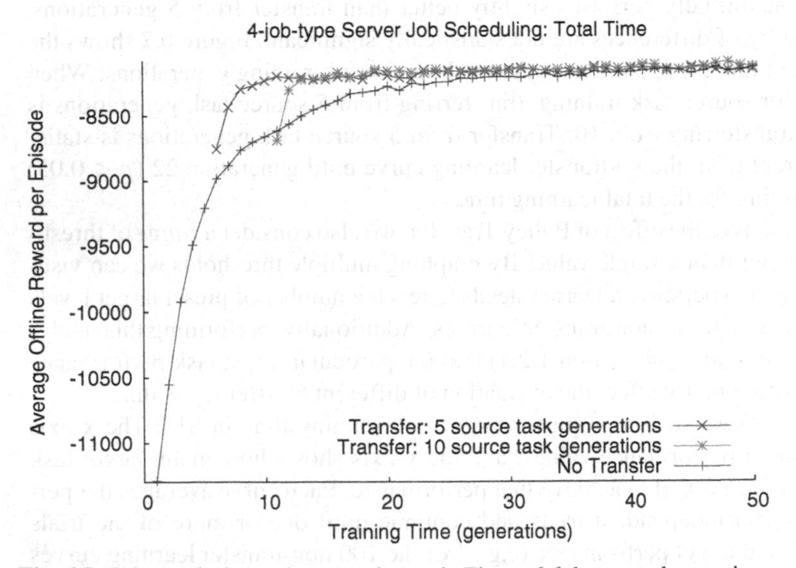

Fig. 6.7 This graph shows the same data as in Figure 6.6, but now the x-axis accounts for the total training time (source generations and target task generations)

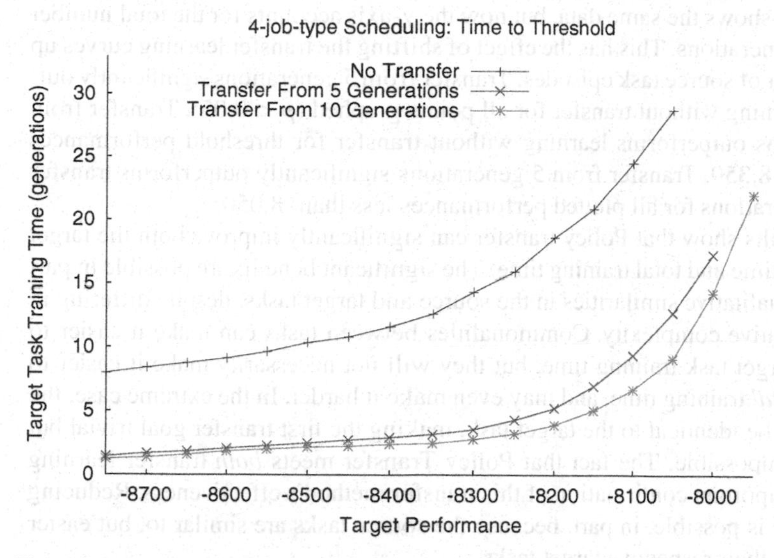

Fig. 6.8 The average number of generations in SJS needed to attain a target performance (the average reward per episode) is successfully reduced via Policy Transfer

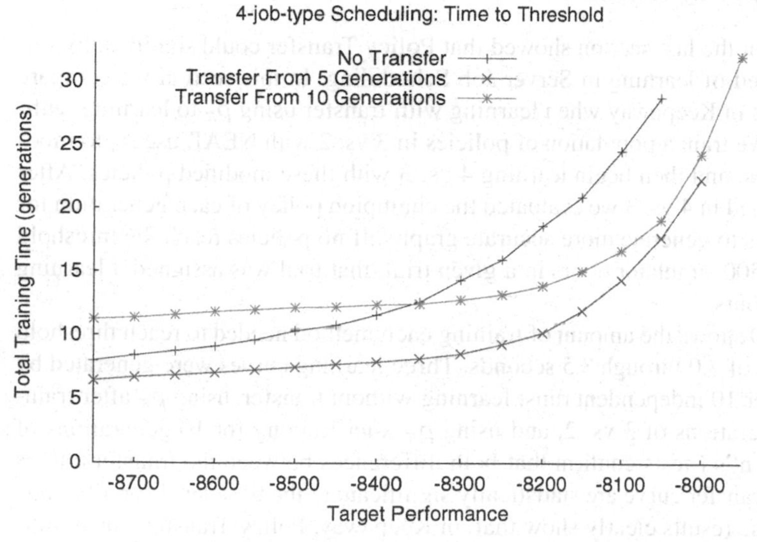

Fig. 6.9 This graph plots the same data as in Figure 6.9, but now the y-axis accounts for source task training time as well as the target task training time

generations to achieve the same performance as the transfer learners. Differences between the numbers of target task episodes are significant at the 95% level for transfer and non-transfer curves for all thresholds graphed.

Figure 6.9 shows the same data, but now the y-axis accounts for the total number of training generations. This has the effect of shifting the transfer learning curves up by the number of source task episodes. Transfer from 5 generations significantly outperforms learning without transfer for all points graphed ($p < 0.05$). Transfer from 10 generations outperforms learning without transfer for threshold performances greater than -8,350. Transfer from 5 generations significantly outperforms transfer from 10 generations for all plotted performances less than -8,050.

These results show that Policy transfer can significantly improve both the target task training time and total training time. The significant benefits are possible in part because of qualitative similarities in the source and target tasks, despite differing in S, A, and relative complexity. Commonalities between tasks can make it easier to reduce (1) target task training time, but they will not necessarily make it easier to reduce (2) *total* training time, and may even make it harder. In the extreme case, the source could be identical to the target task, making the first transfer goal trivial but the second impossible. The fact that Policy Transfer meets *both* transfer learning goals is an important confirmation of this transfer method's effectiveness. Reducing the total time is possible, in part, because the source tasks are similar to, but easier to learn than, their respective target tasks.

6.2.2 Keepaway Results

Experiments in the last section showed that Policy Transfer could significantly improve the speed of learning in Server Job Scheduling. In this section we compare learning times in Keepaway when learning with transfer using ρ_π to learning without transfer. We train a population of policies in 3 vs. 2 with NEAT, use ρ_π to modify the policies, and then begin learning 4 vs. 3 with these modified policies. After learning finished in 4 vs. 3 we evaluated the champion policy of each generation for 1,000 episodes to generate more accurate graphs. If no policies reach the threshold value within 500 simulator hours in a given trial, that trial was assigned a learning time of 500 hours.

Figure 6.10 shows the amount of training each method needed to reach threshold performances of 7.0 through 8.5 seconds. Three learning curves were generated by averaging over 10 independent runs: learning without transfer, using ρ_π after training for 5 generations of 3 vs. 2, and using ρ_π after training for 10 generations of 3 vs. 2. Student's t-tests confirm that both differences between the transfer curves and the non-transfer curve are statistically significant at the 95% level for all points graphed. These results clearly show that, in Keepaway, Policy Transfer can significantly reduce learning times in the target task.

When considering the total training time, learning curves in Figure 6.10 which use transfer are shifted up by the amount of time spent training in the source task. The differences between the total training times with transfer and without transfer are statistically significant for roughly half of the target threshold times shown in Figure 6.10; the benefit from transfer was greater for higher target thresholds.

We hypothesize that transfer from 5 source task generations outperforms transfer from 10 source task generations in both domains due to two factors. First,

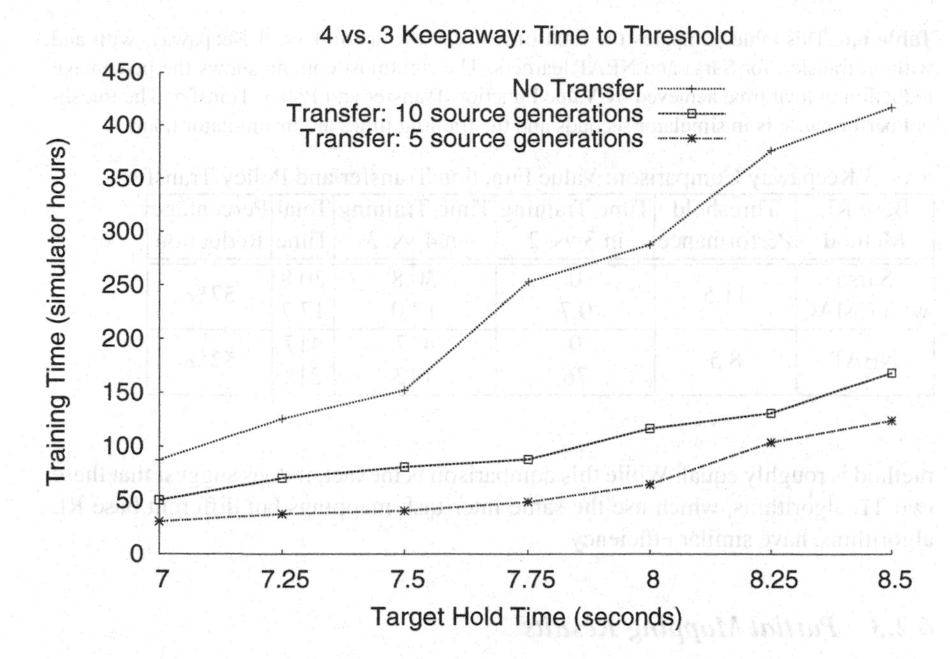

Fig. 6.10 Policy Transfer successfully reduces the average target task training time needed to reach a given performance level, relative to learning without transfer

source task networks trained for 10 generations have more links and nodes than those trained for 5 generations, and more complex networks are likely to learn more slowly than simple networks. Second, training for more time in the source task may lead to overfitting, causing a target task learner to have to spend more time learning to perform well in the target task.[7]

It is difficult to directly compare these Policy Transfer results with the Value Function Transfer results because the players plateau at different levels (the majority of 4 vs. 3 NEAT players reach a performance of 8.5 seconds, while the majority of 4 vs. 3 CMAC players reach a performance of 11.5 seconds). We therefore compare the relative improvements in total performance for the highest thresholds that each set of players consistently achieved.[8] Table 6.6 compares the two transfer methods and shows that the percentage of total time saved by using each transfer

[7] One way to test this hypothesis would be to freeze the source task network topology while training, which would help distinguish between learning by changing weights and learning by adding nodes to the network. If the weights are being overfit, transfer after 10 source task generations would be worse than transfer after 5 source task generations, even if the topology was frozen. If the topology was growing too quickly in the source task, freezing the source task topology would improve the performance of the players transferring from 10 source task generations, relative to those transferring from 5 source task generations.

[8] It is not useful to test the 4 vs. 3 CMAC players with a threshold performance of 8.5 seconds because many of the transfer trials begin learning with a performance higher than 8.5.

Table 6.6 This table compares the total time needed to learn 4 vs. 3 Keepaway, with and without transfer, for Sarsa and NEAT learners. The rightmost column shows the percentage reduction in total time achieved by Value Function Transfer and Policy Transfer. The threshold performance is in simulator seconds and the training times are in simulator hours.

4 vs. 3 Keepaway Comparison: Value Function Transfer and Policy Transfer

Base RL Method	Threshold Performance	Time Training in 3 vs. 2	Time Training in 4 vs. 3	Total Time	Percentage Reduction
Sarsa with CMAC	11.5	0	30.8	30.8	**57%**
		0.7	17.0	17.7	
NEAT	8.5	0	417	417	**52%**
		76	142	218	

method is roughly equal. While this comparison is inexact, it does suggest that these two TL algorithms, which use the same inter-task mappings but different base RL algorithms, have similar efficiency.

6.2.3 Partial Mapping Results

In some situations, the agent may not have access to a full inter-task mapping. One possible simplification is to assume a *partial inter-task mapping*, which may also be easier for a human to intuit or machine to learn. Furthermore, if the target task has actions or state variables that have no correspondence in the source task, a full mapping may be inappropriate. In this section we define partial mappings for the SJS and Keepaway domains and empirically demonstrate that they may be used for successful transfer in conjunction with Policy Transfer.

The full inter-task mapping from the 4-job-type task to the 2-job-type task mapped each target task job type to a source task job type. The incomplete inter-task mappings are defined so that job types 1 and 2 in the target task are mapped to job types 1 and 2 in the source task. The target task's novel job types, 3 and 4, are not assigned any mapping to the source task. Using this mapping, we can construct $\chi_{P,A}$ and $\chi_{P,A}$, where the P subscript denotes "partial." When these incomplete mappings are used for Policy Transfer, the weights in the target task corresponding to the two mapped job types are set via transfer, but nodes corresponding to the two unmapped job types in the target task are initially only connected by links with randomized weights.

When transferring from 3 vs. 2 to 4 vs. 3 Keepaway, we define $\chi_{P,X}$ to be the same as before (see Table 5.1), except for the mapping from novel state variables in 4 vs. 3, which are undefined. For example, *Distance to second closest keeper* in 4 vs. 3 still maps to *Distance to second closest keeper* in 3 vs. 2, but $\chi_{P,X}$(*Distance to third closest keeper*) is undefined. Likewise, $\chi_{P,A}$ is the same as χ_A, the full inter-task mapping, except that $\chi_{P,A}$ (`Pass to third closest keeper`) is undefined.

The partial inter-task mappings in the SJS and Keepaway domains can now be used to construct ψ_P and ρ_P, as was done in the previous section with the full mappings. Figure 6.11 shows the target time results in the SJS domain. Using partial

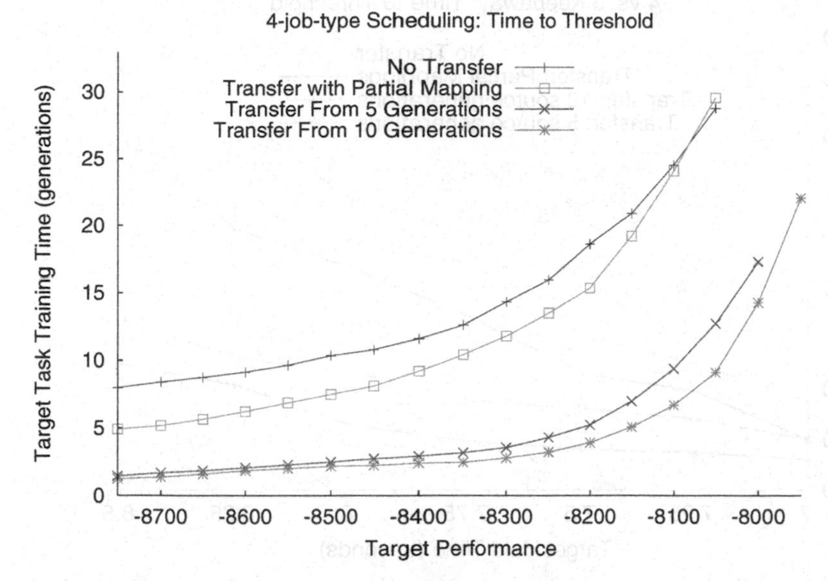

Fig. 6.11 The average number of generations in SJS needed to attain a target performance level (the average reward per episode) is successfully reduced via Policy Transfer. Using partial mappings (after learning for 5 source task generations) outperforms learning without transfer, but underperforms using the full inter-task mappings. The No Transfer curve and the transfer with the full mappings from 5 and 10 source task generations curves are replicated from Table 6.8.

mappings to transfer from 5 source task generations successfully reduces the target task learning time, relative to learning without transfer, for threshold performances less than -8,150 (Student's t-tests: $p < 0.05$). Using the partial mappings also reduces the total time for threshold performances less than -8,250 (not shown).

Figure 6.12 shows results from Keepaway. Over half of the target task time differences between the no transfer and transfer from 5 generations using partial mappings are significant ($p < 0.05$). In both domains, using the partial mappings to transfer from 5 source task episodes does not perform as well as using the full mappings, but Policy Transfer with the partial mappings does outperform learning without transfer.

Taken as a whole, these results show that Policy Transfer can successfully transfer knowledge so that both the target task training time and total training time are reduced. Transfer with partial mappings enables faster learning in the target task than when training without transfer in both domains, but the full inter-task mappings are even more beneficial. This result suggests it is most effective to formulate a full transfer functional between all state variables and actions in the two tasks, but when one is not available, partial inter-task mappings can still successfully improve learning.

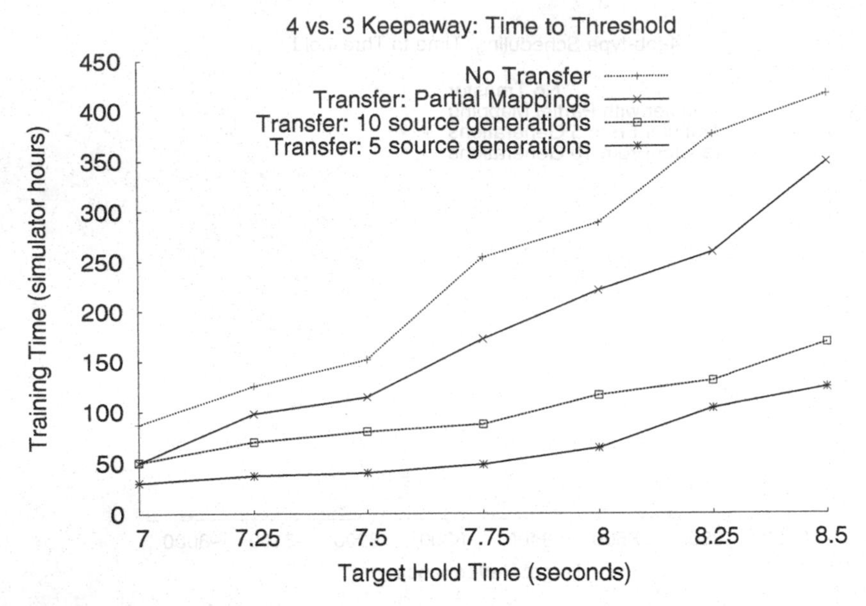

Fig. 6.12 This graph shows the number of generations of target task training required to reach various threshold performance levels in 4 vs. 3 Keepaway with NEAT learners. Partial inter-task mappings (transferring policies learned after 5 source task generations) can successfully reduce target task training time, but are not as effective as the full inter-task mappings. All curves, except for "Transfer: Partial Mappings," are replicated from Figure 6.10.

6.3 Chapter Summary

This chapter has introduced two methods which utilize inter-task mappings to transfer between tasks with different state variables and actions. Q-Value Reuse directly copies a learned action-value function and allows a target task learner to read from it, but not write values into it. Policy transfer shows that a source task population of policies can be modified so that they are a good starting point for a target task learner. In the following chapter we introduce three transfer methods which allow the source task agent and target task agent to use entirely different learning methods, which is not possible when using the TL methods presented thus far.

Chapter 7
Transfer between Different Reinforcement Learning Methods

The previous two chapters introduced three different methods that use inter-task mappings to transfer between tasks with different state variables and actions, but all methods required agents in the source task and target task to use the same type of underlying RL method.

1. Value Function Transfer (Section 5.2) used an action-value function learned in the source task to initialize an action-value function in the target task, with the requirement that both source and target task agents use value-function learning, such as Q-Learning or Sarsa.
2. Q-Value Reuse (Section 6.1) also required TD learners in the source and target task, but copied an entire Q-value function, rather than using it to initialize a target task's action-value function. Thus the target task agent must use value function learning.
3. Policy transfer (Section 6.2) transfers between policy search methods which use neural network action selectors.

In this chapter, we introduce three additional transfer methods which do not require the source and target tasks to be learned by the same type of RL algorithm. Previous work (c.f., [Taylor et al. (2006)]) has shown that characteristics of a particular task may favor one type of RL algorithm over another. If one can determine what type of RL algorithm would be best for a given target task, a TL method would, ideally, be flexible enough to reuse knowledge from a source task, even were that source task learned by a different type of algorithm.

Section 7.1 introduces TIMBREL [Taylor et al. (2008c)], a method that transfers observed instances between tasks. (Recall that we use *instance* to refer to an experienced $\langle s,a,r,s' \rangle$ tuple.) This method utilizes inter-task mappings to directly transfer experienced instances between tasks. The instances are used to construct an initial model of T and R (the transition and reward functions) in the target task, which can significantly reduce the amount of experience needed to learn in the target task. Although our experiments use an instance-based RL algorithm in both the source task and target task (Fitted R-MAX, as discussed in Section 2.3.3), any RL algorithm (such as Sarsa or NEAT) could gather instances in the source task to enable beneficial transfer.

M.E. Taylor: Transfer in Reinforcement Learning Domains, SCI 216, pp. 139–179.
springerlink.com

Next, *Rule Transfer* [Taylor and Stone (2007a)] is introduced in Section 7.2. The key difference between this method and others in this monograph is that a higher-level abstraction is transferred between tasks, rather than recorded instances, full policies, or action-value functions. After the source task is learned using some RL algorithm, the agent records instances in the source task. It then uses a rule-learning algorithm to extract rules which describe the policy (i.e., "If the state is s, then take action a"). These rules are used in the target task to increase the speed of learning with Sarsa, relative to not using transfer. As in TIMBREL, instances are used, but now higher-level rules are the information transferred. Additionally, the transferred rules are used directly as an initial control policy, whereas when instances are directly transferred, they are used to help improve learning speed instead of to direct the actions of the target task agent.

Section 7.3 discusses two types of *representation transfer* [Taylor and Stone (2007b)]. The goal of representation transfer is broader than other transfer algorithms in this monograph. In addition to transfer between tasks, the goals of representation transfer are also to enable transfer between different:

- Function approximator parameterizations (e.g., adding or removing state variables from a CMAC)
- Function approximators (e.g., change from an ANN to a RBF)
- Learning methods (e.g., change from Sarsa to policy search)

In all of these cases, the goal of representation transfer is to reuse knowledge between representations that the target representation can be learned faster, relative to not using transfer.

Lastly, Section 7.4 summarizes the six TL methods in this monograph, all of which can utilize the same inter-task mappings. In addition to providing guidelines about when each method would be most appropriate, we provide a chart summarizing the experiments, RL algorithms, and function approximators used.

7.1 TIMBREL: Instance-Based Transfer

Model-free algorithms such as Q-Learning and Sarsa learn to predict the utility of each action in different situations but they do not learn the effects of actions. In contrast, model-based (or model-learning) methods, such as Dyna-Q [Sutton and Barto (1998)], PEGASUS [Ng and Jordan (2000)], R-MAX [Brafman and Tennenholtz (2002)], and Fitted R-MAX [Jong and Stone (2007)], use their experience to learn an internal model of how the actions affect the agent and its environment, an approach empirically shown to often be more sample efficient. Such a model can be used in conjunction with *dynamic programming* [Bellman (1957)] to perform off-line planning, often enabling performance superior to model-free methods because better performance can be achieved with fewer environmental samples. Building these models may be computationally intensive, but using CPU cycles to reduce data collection time is a highly favorable tradeoff in many domains, such as in physically embodied agents. In order to further reduce sample complexity, this section introduces *Transferring Instances for Model-Based REinforcement Learning* (TIMBREL), a novel approach to combining TL with model-based RL.

The key insight behind TIMBREL is that data gathered in a source task can be used to build beneficial models in a target task. Data is first recorded in a source task, transformed so that it applies to a target task, and then used by the target task learner as it builds its model. In this section we utilize Fitted R-MAX, an instance based model-learning algorithm, and show how TIMBREL can help learn a target task model by using source task data. TIMBREL combines the benefits of transfer with those of model-based learning to reduce sample complexity. We fully implement and test our method in a set of Mountain Car tasks, demonstrating that transfer can significantly reduce the sample complexity of learning.

In principle, the core TIMBREL algorithm could be used with multiple instance-based model-learning algorithms, but we leave such extensions to future work. Compact, parameterized models can also be learned by some RL methods; it should be possible to transfer such a model (such as a regression model) directly between tasks, but TIMBREL does not directly address this. The experiments use TIMBREL by applying it to Fitted R-MAX (detailed in Section 7.1.2), as it can both learn in continuous state spaces and has had significant empirical success [Jong and Stone (2007)]. Results demonstrate that TIMBREL works in continuous state spaces, as well as between tasks with different state variables and action spaces.

7.1.1 Model Transfer

This section provides an overview of TIMBREL. In order to transfer a model, our method takes the novel approach of transferring observed instances from the source task. The tuples, in the form (s,a,r,s'), describe experience the source task agent gathered while interacting with its environment. One advantage of this approach, compared to transferring an action-value function or a full environmental model (e.g., the transition function), is that the source task agent is not tied to a particular learning algorithm or representation — whatever RL algorithm that learns in the target task will necessarily have to interact with the task and collect experience. This flexibility allows a source task algorithm to be selected based on characteristics of the task, rather than on demands of the transfer algorithm.

To translate a source task tuple into an appropriate target task tuple we again utilize inter-task mappings. When learning in the target task, TIMBREL specifies when to use source task instances to help construct a model of the target task. Briefly, when insufficient target task data exists to estimate the effect of a particular (x,a) pair, instances from the source task are transformed via an inter-task mapping, and are then treated as a previously observed transition in the target task model. The TIMBREL method is summarized in Algorithm 10.

Notice that TIMBREL performs the translation of data from the source task to the target task (line 10) on-line while learning the target task. In Section 7.1.2 we detail how the current state x that is being approximated will affect how the source task sample is translated in our particular task domain. Only transferring instances that will be immediately used thus limits necessary computation. Furthermore, this

Algorithm 10. TIMBREL

1: Learn in the source task, recording (s,a,r,s') transitions.
2: Provide recorded transitions to the target task agent.
3: **while** training in the target task **do**
4: **if** the model-based RL algorithm is unable to accurately estimate some $T(\mathrm{x},a)$ or $R(\mathrm{x},a)$ **then**
5: **while** $T(\mathrm{x},a)$ or $R(\mathrm{x},a)$ does not have sufficient data **do**
6: Locate 1 or more saved instances that, according to the inter-task mappings, are near the current x,a to be estimated.
7: **if** no such unused source task instances exist **then**
8: **exit** the inner while loop
9: Use x, a, the saved source task instance, and the inter-task mappings to translate the saved instance into one appropriate to the target task.
10: Add the transformed instance to the current model for x, a.

method minimizes the number of source instances that must be reasoned over in the target task model by only transferring source task data that will be immediately used.

7.1.2 *Implementing TIMBREL in Mountain Car*

In this section we detail how TIMBREL is used to transfer between tasks in the Mountain Car domain when using Fitted R-MAX as the base RL algorithm. The core TIMBREL result in this monograph is to demonstrate transfer between the Low Power 2D Mountain Car task and the Low Power 3D Mountain Car task. Experiments are summarized in Table 7.1. After learning the 2D task, TIMBREL must be provided an inter-task mapping between the two tasks. We use the same inter-task mappings for Mountain Car as when testing Q-Value Reuse (see Section 6.1.1 and Table 6.2). Note that the state variable mapping is defined so that either the target task state variables (x and $\dot{x}$) *or* (y and $\dot{y}$) are mapped into the source task. As we will discuss, the unmapped target task state variables are set by the state variables' values in the state x that we wish to approximate.

As discussed in Section 2.3.3, Fitted R-MAX approximates transitions from a set of sample states $\mathrm{x} \in X$ for all actions. When the agent initially encounters the target task, no target task instances are available to approximate T. Without transfer, Fitted R-MAX would be unable to approximate $T(\mathrm{x}_{target}, a_{target})$ for any x and would set the value of $Q(s_{target}, a_{target})$ to an optimistic value (R_{max}) to encourage exploration. Instead, TIMBREL is used to transfer instances from the source task to help approximate $T(\mathrm{x}_{target}, a_{target})$.

TIMBREL is given a set of source task instances and inter-task mappings as inputs and must construct one or more target task tuples, $(s_{target}, a_{target}, r, s'_{target})$, to help approximate $T(\mathrm{x}_{target}, a_{target})$. The goal of transfer is to find some source task tuple $(s_{source}, a_{source}, r, s'_{source})$ where $a_{source} = \chi_A(a_{target})$ and s_{source} is "near" $\chi_X(s_{target})$

Table 7.1 This table summarizes TIMBREL experiments in this section of the monograph

Method Name: TIMBREL			
Scenario: This TL method is applicable whenever the target task agent uses instance-based RL. The source task agent may use any RL method, but in our experiments we learn with Fitted R-MAX in both the source task and target task.			
Source Task	**Target Task**	**Function Approximator**	**Section**
Low Power 2D Mountain Car	Low Power 3D Mountain Car	Instances	7.1.3
High Power 2D Mountain Car	Low Power 3D Mountain Car	Instances	7.1.3
No Goal 2D Mountain Car	Low Power 3D Mountain Car	Instances	7.1.3

(line 6 in Algorithm 10). Once we identify such a source task tuple, we can then use χ^{-1} to convert the tuple into a transition appropriate for the target task (line 9), and add it to the data approximating T (line 10).

As an illustrative example, consider the case when the agent wants to approximate $T(\mathtt{x}_{target}, a_{target})$, where

$$\mathtt{x}_{target} = \langle x_{target}, y_{target}, \dot{x}_{target}, \dot{y}_{target} \rangle = \langle -0.6, -0.2, 0, 0.1 \rangle$$

and $a_{target} = \mathtt{East}$. TIMBREL considers source task transitions that contain the action `Right` (i.e., $\chi_A(\mathtt{East})$). χ_S is defined so that either the x or y state variables can be mapped from the target task to the source task, which means that we should consider two transitions selected from the source task instances. The first tuple is selected to minimize the Euclidean distances

$$\sqrt{(x_{target} - x_{source})^2 - (\dot{x}_{target} - \dot{x}_{source})^2},$$

where each state variable is scaled by its range. The second tuple is chosen to minimize

$$\sqrt{(y_{target} - x_{source})^2 - (\dot{y}_{target} - \dot{x}_{source})^2}.$$

Continuing the example, suppose that the first source task tuple selected was

$$(\langle -0.61, 0.01 \rangle, \text{Right}, -1, \langle -0.59, 0.02 \rangle).$$

If the inter-task mapping defined mappings for the x and y state variables simultaneously, the inverse inter-task mapping *could* be used to convert the tuple into

$$(\langle -0.61, -0.61, 0.01, 0.01 \rangle, \mathtt{East}, -1, \langle -0.59, -0.59, 0.02, 0.02 \rangle).$$

However, this point is not near the current x_{target} we wish to approximate. Instead, we recognize that this sample was selected from the source task to be near to x_{target} and $\dot{x}_{target}$, and transform the tuple, assuming that y_{target} and $\dot{y}_{target}$ are kept constant. With this assumption, we form the target task tuple

$$(\langle -0.61, y_{target}, 0.01, \dot{y}_{target}\rangle, \texttt{East}, -1, \langle -0.59, y_{target}, 0.2, \dot{y}_{target}\rangle) = (\langle -0.61, -0.2, 0.01, 0\rangle, \texttt{East}, -1, \langle -0.59, -0.2, 0.02, 0\rangle).$$

The analogous step is then performed for the second selected source task tuple; we transform the source task tuple with $\mathcal{X}$ while assuming that x_{target} and $\dot{x}_{target}$ are held constant. Finally, both transferred instances are added to the approximation of $T(\mathrm{x}, a)$.

TIMBREL thus transfers pairs of source task instances to help approximate the transition function. Other model-learning methods may need constructed trajectories instead of individual instances, but TIMBREL is able to *generate* trajectories as well. Over time, the learner will approximate $T(\mathrm{x}_{target}, a_{target})$ for different values of (x, a) in order to construct a model for the target task environment. Any model produced via transfer may be incorrect, depending on how representative the saved source task instances are of the target task (as modified by $\mathcal{X}$). However, our experiments demonstrate that using transferred data may allow a model learner to produce a model that is more accurate than if the source data were ignored.

As discussed in Section 2.3.3, Fitted R-MAX uses the distance between instances and x to calculate instance weights. When an instance is used to approximate x, that instance's weight is added to the total weight of the approximation. If the total weight for an approximation does not reach a threshold value of 1.0, an optimistic value (R_{max}) is used because not enough data exists for an accurate approximation. When using TIMBREL, the same calculation is performed, but now instances from both the source task and target task can be used.

As the agent interacts with the target task, more transitions are recorded and the approximations of the transition function at different (x, a) pairs need to be recalculated based on the new information. Each time an approximation needs to be recomputed, Fitted R-MAX first attempts to use only target task data. If the number of instances available (where instances are weighted by their distance from x) does not exceed the total weight threshold, source task data is transferred to allow an approximation of $T(\mathrm{x}_{target}, a_{target})$. This process is equivalent to removing transferred source task data from the model as more target task data is observed, allowing the model's accuracy to improve over time. Again, if the total weight from source task and target tasks instances for an approximated x does not reach 1.0, R_{max} is assigned to the model as the reward for reaching x.

As a final implementation note, consider what happens when some x maps to an s_{source} that is not near any experienced source task data. If there are no source task transitions near s_{source}, it is possible that using all available source task data will not produce an accurate approximation (recall that instance weights are proportional to the square of the distance from the instance to x). To avoid a significant increase in computation complexity with limited improvement in approximating T, we imposed a limit of 20 source task tuples when approximating a particular point (line 5). This threshold serves a similar purpose as the 10% cumulative weight threshold (the "minimum fraction" parameter) discussed in Section 2.3.3.

7.1.3 TIMBREL Transfer Experiments

In order to test the efficacy of transfer, first we conducted an experiment to measure the learning speed of Fitted R-MAX in the Mountain Car domain both with and without TIMBREL. Roughly 50 different sets of Fitted R-MAX parameters were used in preliminary experiments to select the best settings for learning the 3D task without transfer (as discussed in Section 2.3.3). We ran 12 trials for 4,000 episodes and found that 10 out of 12 trials were able to converge to a policy that found the goal area. Recall that Fitted R-MAX is not guaranteed to converge to an optimal policy because it depends on approximation in a continuous state space.

To transfer from the Low Power 2D Mountain Car task into the more complex Low Power 3D Mountain Car, we first allow 12 Fitted R-MAX agents to train for 100 episodes each in the 2D task while recording all observed (s,a,r,s') transitions.[1] We then used TIMBREL to train agents in the target task for 1,000 episodes. 12 out of 12 trials converged to a policy that found the goal area.

After learning, we averaged over all non-transfer and transfer learning trials. For clarity, we also smoothed the curves by averaging over a 10 episode window. Figure 7.1(a) shows the first 1000 episodes of training (running the experiments longer than 1,000 episodes did not significantly improve the policy, as suggested by Figure 4.6). Student's t-tests determined that all the differences in the averages were statistically significant ($p < 0.05$), with the exception of the first data point (episode 9). This result confirms that transfer can significantly improve the performance of agents in the 3D Mountain Car task.

We hypothesize that the U-shaped transfer learning curve is caused by a group of agents that find an initial path to the goal, spend some number of episodes exploring to find a faster path to the goal, and ultimately return to the original policy (see Figure 7.1(b)). In addition to improved initial performance, the asymptotic performance is improved, in part because some of the non-transfer tasks failed to successfully locate the goal. The difference in success rates (10 of 12 trials reaching the goal vs. 12 of 12) suggests that transfer may make difficult problems more tractable.

TIMBREL, and its implementation, were designed to minimize sample complexity. However, it is worth noting that there is a significant difference in the computational complexity of the transfer and non-transfer methods. Every time the transfer agent needs to use source task data to estimate T, it must locate the most relevant data and then insert it into the model. Additionally, the transfer agent has much more data available initially and adding additional data to the model is significantly

[1] We experimented with roughly 10 different parameter settings for Fitted R-MAX in the Low Power 2D task. Every episode lasts 500 time steps if the goal is not found and the 2D goal state can be reached in roughly 150 time steps. When learning Low Power 2D Mountain Car, the agent experienced an average of 24,480 source task transitions during the 100 source task episodes.

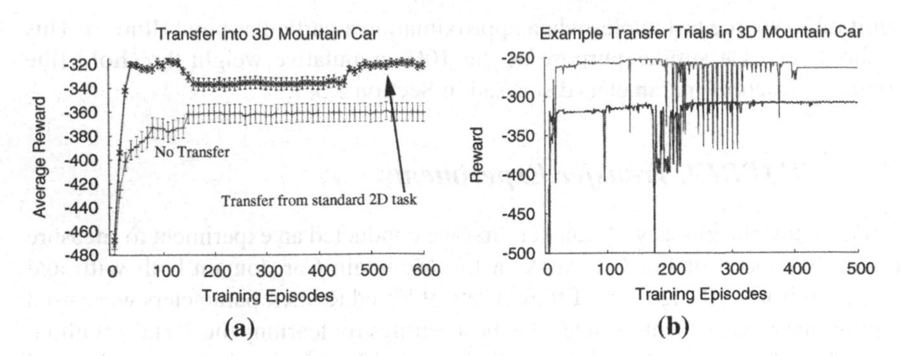

Fig. 7.1 **(a)** TIMBREL significantly improves the speed of Fitted R-MAX on the 3D Mountain Car task. The average performance is plotted every 10 episodes along with the standard error. **(b)** Two example Fitted R-MAX learning curves show that the on-line performance can vary significantly, sometimes resulting in a U-shaped learning curve, visible in (a) when multiple trials are averaged.

slower than the non-transfer agent. These factors cause the transfer learning trials to take roughly twice as much wall clock time as the non-transfer trials. While further code optimizations could be added, using the additional transferred data will always slow down the agent's computation, relative to an agent that is not using transfer, but executes the same number of actions.

Our second experiment examines how the amount of recorded source task data affects transfer. One hypothesis was that more tuples in the source task would equate to higher performance in the target task, because the target task agent would have more data to draw from, and thus would be better able to approximate any given $T(\mathrm{x},a)$.

Our second experiment trained source task agents in the Low Power 2D task for 5, 10, and 20 episodes. Figure 7.2 shows that transfer from 20 source task episodes is similar to using 100 source task episodes and performs statistically better than no transfer at the 95% level for 98 of the 100 points graphed. While transfer performance degrades for trials that use 10 and 5 source task episodes, both trials do show a statistically significant boost to the agents' *initial* learning performance. This result demonstrates that a significant amount of information can be learned in just a few source task episodes; the source task is less complex than the target task and thus a short amount of time spent learning in the source may have a large impact on the target task performance.

Recall the Mountain Car has a reward of -1 on each time step. The agent learns to reach the goal area because transitioning into this area ends the episode and the steady stream of negative reward. The third experiment uses the No Goal 2D task (introduced in Section 4.1.1) as a source task to examine how changing the reward function in the source task affects transfer. When training in the source task, every episode lasted 500 time steps (the maximum number of steps). After learning for 100 episodes in the source task, we transferred into the target task and found that

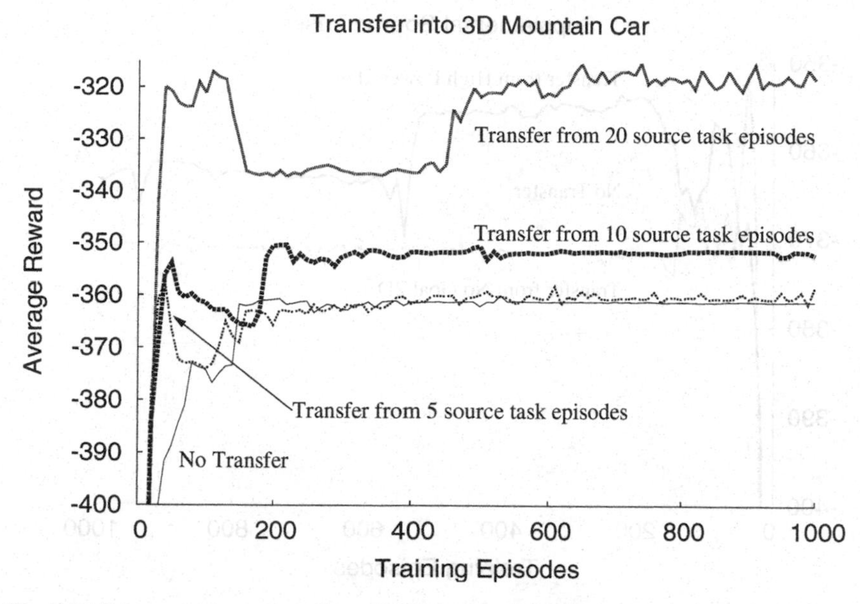

Fig. 7.2 This graph shows the effect of different amounts of source task training. Each learning curve is the average of 12 independent trials.

9 of the 12 trials successfully discovered policies to reach the goal area. Figure 7.3 suggests that transfer from a source task policy with a different reward structure can be initially useful (t-tests confirm that transfer outperforms non-transfer for four of the first five points graphed), but the relative performance of the non-transfer trials soon outperform that of learning with transfer.

Our fourth experiment uses the High Power 2D task as a source task. We again record 100 episodes worth of data with source task learners and use TIMBREL to transfer into Low Power 3D Mountain Car. However, because the source task uses a car with a motor more than twice as powerful as in the 3D task, the transition function learned in the source task is less useful to the agent in the target task than when transferring from the Low Power 2D task. 9 of the 12 target task trials successfully converged to a policy that reached the goal. Figure 7.3 shows that the average performance of a transfer learner using the High Power 2D task as a source performs worse than when transferring from the Low Power 2D task (shown in Figures 7.1(a) and 7.2). Although t-tests show that there is a statistically significant improvement at the beginning of learning, the transfer and non-transfer curves quickly become statistically indistinct with more target task training.

Figure 7.3 highlights an important drawback of transfer learning. Transfer efficacy is often affected by the similarity of source tasks and target tasks, and in some circumstances transfer may not help the learner. Indeed, if T or R in the source and target tasks are too dissimilar, transfer may actually cause the learner to learn more slowly than if it had not used transfer. While there is not yet a general solution

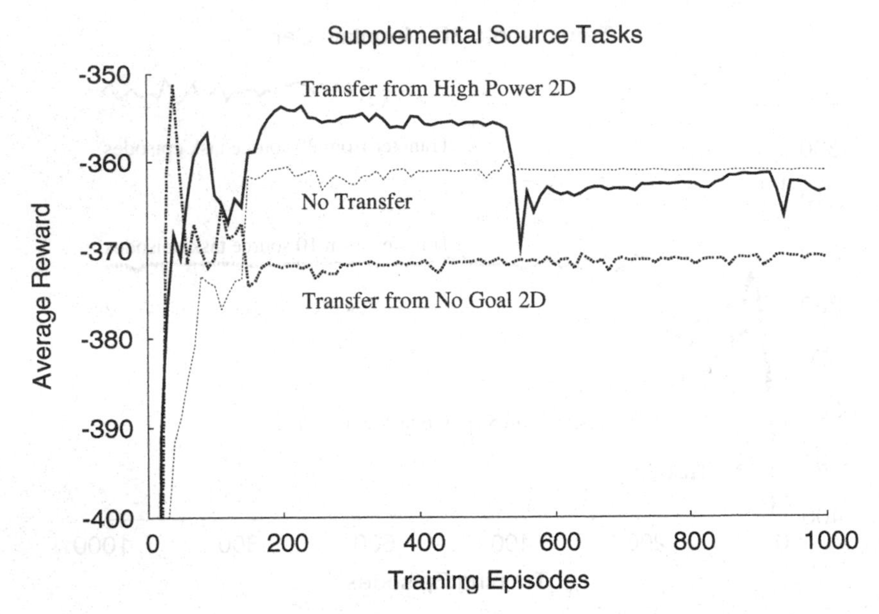

Fig. 7.3 Transfer from a 2D Mountain Car task that has no goal state or from a 2D Mountain Car with significantly stronger acceleration produces statistically significant improvements at the beginning of learning when compare to learning without transfer. However, this relative advantage is lost as agents in the target Low Power 3D task gain more experience.

to avoiding *negative transfer*, other recent results (Section 8.2) suggest that the "relatedness" of tasks may be measured empirically, and may guide learners when deciding whether or not to transfer.

7.1.4 TIMBREL Summary

This section has introduced TIMBREL, a transfer method compatible with model-based reinforcement learning. We demonstrate that when learning 3D Mountain Car with Fitted R-MAX, TIMBREL can significantly reduce the sample complexity. Furthermore, experiments demonstrate how transfer is affected by changes to the learned source task's reward and transfer functions as they become less similar to the target task's reward and transfer functions.

7.2 Transfer via Rules

This section introduces *Rule Transfer*, a novel domain-independent RL transfer method. Similar to TIMBREL, Rule Transfer records instance in a source task to learn a target task faster. However, rather than transferring the instance directly, we first learn production rules (henceforth *rules*) to summarize the source task policy.

Rule Transfer then uses inter-task mappings to transform the rules so that they can apply to the target task, even when the target task agent has a different internal representation from the source task agent. Thus one agent may train very quickly with a simple internal representation in the source task, but a second, more complex agent in the target task can still benefit from transfer.

After the Rule Transfer algorithm is described, Section 7.2.1 evaluates three different possible rule utilization schemes for Rule Transfer in Keepaway. We then empirically show that cross-domain transfer can effectively improve the speed of learning on tasks drawn from different domains in Section 7.2.2 (namely, transfer from Ringworld to Keepaway and from Knight Joust to Keepaway). The last experiment (Section 7.2.3) uses Rule Transfer to learn 4 vs. 3 Keepaway using rules learned in 3 vs. 2 Keepaway. Section 7.2.4 concludes the discussion of Rule Transfer.

Figure 7.4 and the following list of steps describes Rule Transfer:

1. **Learn a policy ($\pi : S \mapsto A$) in the source task.** After training has finished, or during the final training episodes, the agent records some number of interactions with the environment in the form of (s, a) pairs while following the learned policy.
2. **Learn a decision list ($D_s : S \mapsto A$) that summarizes the source policy.** After the data is collected, a rule learner is used to summarize s, a pairs. These rules can be used to directly control a source task agent with approximately the same policy as was learned in the source task.
3. **Modify the decision list for use in the target task** (`Translate`$(D_s) \rightarrow D_t$). To allow the learned decision list to be applied by an agent in a target task (that

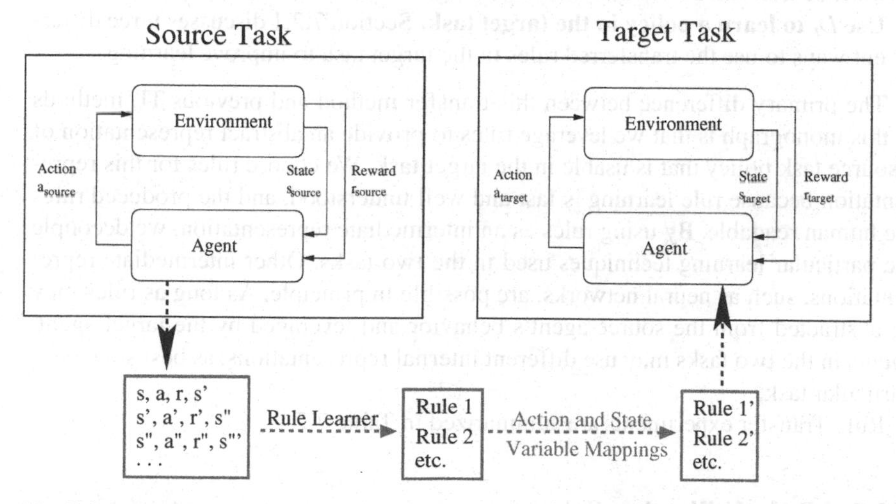

Fig. 7.4 Rule Transfer has four main steps. First, after learning, the source task agent saves a number of experienced instances. Second, propositional rules are learned to summarize saved source task instances. Third, the rules are transformed via inter-task mappings. Fourth, the rules are used in the target task to speed up learning.

Table 7.2 This table summarizes the Rule Transfer experiments in this section of the monograph

Method Name: Rule Transfer			
Scenario: Rule Transfer is applicable when the target task agent uses value function learning. Rules allow agents to transfer between different RL algorithms and function approximators. All experiments use Sarsa in the source task and in the target task.			
Source Task	**Target Task**	**Function Approximator**	**Section**
3 vs. 2 Keepaway	3 vs. 2 Keepaway	RBF	7.2.1.1
Ringworld	3 vs. 2 Keepaway	Tabular and RBF	7.2.2.1
Knight Joust	3 vs. 2 Keepaway	Tabular and RBF	7.2.2.2
Knight Joust	4 vs. 3 Keepaway	Tabular and CMAC	7.2.2.2
3 vs. 2 Keepaway	4 vs. 3 Keepaway	CMAC	7.2.3

has different state variables and actions from the source task), the decision list must be translated with inter-task mappings before it can be used.

4. **Use D_t to learn a policy in the target task.** Section 7.2.1 discusses three different ways to use the transferred rules in the target task to improve learning.

The primary difference between this transfer method and previous TL methods in this monograph is that we leverage rules to provide an abstract representation of a source task policy that is usable in the target task. We choose rules for this representation because rule learning is fast and well understood, and the produced rules are human readable. By using rules as an intermediate representation, we decouple the particular learning techniques used in the two tasks. Other intermediate representations, such as neural networks, are possible in principle. As long as rules may be abstracted from the source agent's behavior and leveraged by the target agent, agents in the two tasks may use different internal representations, as best suits their particular task.

Rule Transfer experiments are summarized in Table 7.2.

7.2.1 Rule Utilization Schemes

If the target task has different state variables or actions than the source task, or they have different semantic meanings in the two tasks, an agent could not directly apply a learned decision list from the source task because the preconditions for the rules and/or actions recommended would be inapplicable. The function `Translate()` (step #3 above) procedurally modifies the source task decision list via inter-task mappings so that it can apply to a given target task. If source task state variables or actions had no correspondence in the target task (i.e., only partial inter-task mappings existed), the affected preconditions or rules would be removed from the translated decision list.

If the production rules were strictly followed by the target task agent, the agent may receive an initial benefit, relative to not using transfer. However, no learning would be possible unless the agent could deviate from the transferred rules. To make Rule Transfer effective, we thus treat the translated decision list as *advice*. The agent may receive an initial jumpstart by strictly following the decision list, but then should refine its policy as it gathers more experience in the target task. This section introduces three advice utilization schemes (summarized in Figure 7.5). The first method applies only if the target task learner is using a value-function approximation method, but the second and third may be used in conjunction with other RL learning methods, possibly with minor modifications.

The *Value Bonus* rule utilization scheme uses the transferred decision list, D_t, to determine which target task action the decision list would recommend in the current state. The computed Q-value of this recommended action then receives a "bonus" so that it is increased by some constant (which is set empirically, as described in the next section). Actions recommended by D_t are initially more likely to be selected, but the bonus can be overridden through learning. Note that this scheme is similar to adding a shaping reward to actions selected by the translated decision list.

Extra Action adds an action to the target task. When the target task agent selects this pseudo-action, the agent executes the action recommended by D_t in the "true" MDP. The learner treats this pseudo-action no differently than other actions when learning. To bias the learner towards this action, the agent is forced to execute the pseudo-action for a constant number of episodes at the beginning of training in the target task. Afterward (assuming the agent does not use optimistic initialization) the pseudo-action will have higher Q-values than all other actions, which causes the agent to initially perform recommended actions. Over time the agent can learn to override this bias. For instance, in regions of the state space in which the advice is appropriate, the agent will learn to select the pseudo-action, while in other regions of the state space where the advice is non-optimal, the agent must learn to intelligently choose between all the actions.

Extra Variable adds an extra state variable to the target task's state description, creating an augmented MDP, different from that created by the Extra Action method. The extra variable takes on the value of the index for the action recommended by D_t. To assist the agent in learning the importance of this variable, we again initially force the agent to choose the action recommended by D_t. An agent quickly learns the importance of this state variable, but it can still learn to ignore the state variable when the advice is sub-optimal.

Ignore Rules

$Q(x_1, x_2, a_1) = 5$
$Q(x_1, x_2, a_2) = 3$
$Q(x_1, x_2, a_3) = 4$

Value Bonus

$Q(x_1, x_2, a_1) = 5$
$Q(x_1, x_2, a_2) = 3 + 10$
$Q(x_1, x_2, a_3) = 4$

Extra Action

$Q(x_1, x_2, a_1) = 5$
$Q(x_1, x_2, a_2) = 3$
$Q(x_1, x_2, a_3) = 4$
$Q(x_1, x_2, a_4) = 13$

Extra Variable

$Q(x_1, x_2, x_3, a_1) = 5$
$Q(x_1, x_2, x_3, a_2) = 13$
$Q(x_1, x_2, x_3, a_3) = 4$

Fig. 7.5 This figure shows the three different rule utilization schemes described in the text by using an example MDP with two state variables and 3 actions. At top are hypothetical Q-values in a state s when the transferred rules are ignored. Suppose that a transferred decision list suggested that action a_2 should be taken. Value Bonus adds some constant to the action-value of action a_2. Extra Action adds an extra action, a_4, to the MDP. If the agent chooses this action, it will execute the action suggested by the decision list (which is a_2 in this example). Extra Variable adds a state variable, x_3, which takes on the action index suggested by the decision list (which would make $x_3 = 2$ in this example because action 2 is suggested).

7.2.1.1 Testing Rule Utilization in Keepaway

To determine reasonable settings for the different rule utilization methods outlined in Section 7.2.1, we analyze Rule Transfer by using $25m \times 25m$ 3 vs. 2 Keepaway as the source *and* target task.[2] We first train in 3 vs. 2 for five simulator hours (roughly 1,300 episodes). Next, JRip, an implementation of RIPPER [Cohen (1995)] included in Weka [Witten and Frank (2005)], learns a decision list summarizing the source task policy.[3] Lastly, we utilize the decision list in a new instance of Keepaway.

Figure 7.6 shows the performance averaged over 10 learning trials for learning without transfer with a 1000 episode sliding window. A second learning curve shows the performance when always utilizing the rules ("Always Use Rules"). Additionally, the three rule utilization methods are shown. Table 7.3 details the results, which show that while transfer is affected by the relevant rule utilization parameters, each rule utilization method has a wide range of effective parameters. All three methods significantly improve the three measured TL metrics.

[2] In practice, this would not be a useful TL procedure in and of itself. By transferring from a task into the same task, we are able to study parameter settings for Rule Transfer.

[3] RIPPER is a simple propositional rule learner that can learn a decision list. If additional representational power were needed, an ILP rule learner like Aleph [Srinivasan (2001)] could be used, but we found the additional complexity unnecessary.

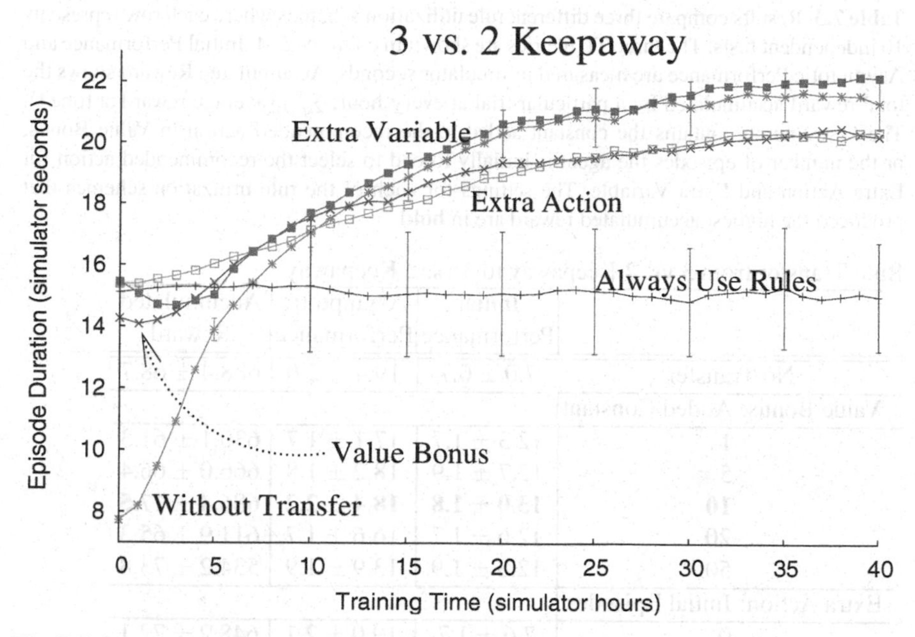

Fig. 7.6 This graph shows the average of 10 independent learning trials for learning without transfer and for using rules after learning for five simulator hours (the source task learning time is not shown): without further learning, with a Value Bonus of +10, with Extra Action after 100 episodes, and with Extra Variable after 100 episodes. Learning without transfer and using rules without learning have standard error bars in 5 hour increments (which are not shown on all lines for visual clarity). These results were collected using version 0.6 of the Keepaway benchmark players.

7.2.2 *Cross-Domain Rule Transfer Results*

As discussed earlier in Section 5.3.4, cross-domain transfer has been a long-term goal of transfer learning because it could allow transfer between significantly different tasks. While the majority of transfer work has focused on reducing training time by transferring from a simple to complex task in a single domain, a (potentially) more powerful way of simplifying a task is to formulate it as an abstraction in a different domain. In this section we show that source tasks drawn from the gridworld domain can significantly improve learning in Keepaway, even though learning in the gridworld tasks take orders of magnitude less wall clock time than learning Keepaway tasks.

In the previous section we determined appropriate advice utilization settings via three transfer metrics. In this section we apply those same settings while using Ringworld and Knight Joust as source tasks and Keepaway as the target. This section demonstrates that transfer from Ringworld is able to significantly improve all three

Table 7.3 Results compare three different rule utilization schemes where each row represents 10 independent tests. The three TL metrics are shown in columns 2–4. Initial Performance and Asymptotic Performance are measured in simulator seconds. Accumulated Reward shows the total reward accumulated by a particular trial at every hour: $\Sigma_{t=0}^{40}$(average reward at time t). The first column contains the constant added to the recommended action in Value Bonus, or the number of episodes the agent is initially forced to select the recommended action for Extra Action and Extra Variable. The settings for each of the rule utilization schemes that produced the highest accumulated reward are in **bold**.

Rule Transfer from 3 vs. 2 Keepaway to 3 vs. 2 Keepaway

	Initial Performance	Asymptotic Performance	Accumulated Reward
No Transfer	7.0 ± 0.7	19.4 ± 2.0	688.4 ± 68.7
Value Bonus: Added Constant			
1	12.3 ± 1.7	17.1 ± 1.7	630.1 ± 61.3
5	12.7 ± 1.9	18.2 ± 1.8	666.0 ± 66.4
10	**13.0 ± 1.8**	**18.4 ± 2.2**	**686.4 ± 77.5**
20	12.6 ± 1.7	16.6 ± 1.7	611.9 ± 65.3
50	12.8 ± 1.9	13.9 ± 1.9	534.2 ± 73.8
Extra Action: Initial Episodes			
0	7.6 ± 1.7	19.0 ± 2.1	648.2 ± 72.1
50	13.8 ± 2.1	18.3 ± 2.0	676.1 ± 75.4
100	**14.0 ± 2.4**	**18.5 ± 2.0**	**688.2 ± 75.7**
250	13.9 ± 2.3	18.4 ± 1.9	675.3 ± 69.0
500	13.7 ± 2.2	18.1 ± 1.9	678.8 ± 72.2
1000	13.4 ± 2.0	17.9 ± 2.1	648.2 ± 72.1
Extra Variable: Initial Episodes			
0	7.0 ± 0.7	20.0 ± 2.0	691.4 ± 68.8
50	13.6 ± 2.0	19.9 ± 2.0	715.9 ± 70.2
100	14.0 ± 2.3	20.1 ± 2.1	726.0 ± 72.4
250	13.7 ± 2.1	19.9 ± 2.1	717.6 ± 74.6
500	**13.6 ± 2.2**	**20.2 ± 2.0**	**729.2 ± 73.8**
1000	13.7 ± 2.4	17.4 ± 6.3	637.4 ± 207.6

transfer metrics, that the Rule Transfer settings are not particularly brittle when transferring from Ringworld, and that transfer from Knight Joust is able to significantly improve jumpstart in Keepaway, even though the tasks are quite different. As discussed in Section 4.4, Ringworld is a fully observable task with a discrete state space and stochastic actions. Knight Joust (Section 4.5), is a fully observable task with a discrete state space and deterministic player actions. In contrast, Keepaway is partially observable, has a continuous state space, stochastic actions, and has multiple learning agents.

Table 7.4 This table describes the inter-task mappings from 3 vs. 2 Keepaway to Ringworld. These two mappings are used by `Translate()` to modify a decision list learned in Ringworld so that it can apply to Keepaway.

Inter-Task Mappings for Keepaway to Ringworld

Keepaway	**Ringworld**
χ_A	
Hold Ball	Stay
$Pass_1$: Pass to K_2	Run_{Near}
$Pass_2$: Pass to K_3	Run_{Far}
χ_X	
$dist(K_1,T_1)$	$dist(P,O)$
$dist(K_1,K_2)$	$dist(P,Target_1)$
$\text{Min}(dist(K_2,T_1), dist(K_2,T_2))$	$dist(Target_1,O)$
$\text{Min}(ang(K_2,K_1,T_1)$ $ang(K_2,K_1,T_2))$	$ang(O,P,Target_1)$
$dist(K_1,K_3)$	$dist(P,Target_2)$
$\text{Min}(dist(K_3,T_1), dist(K_3,T_2))$	$dist(Target_2,O)$
$\text{Min}(ang(K_3,K_1,T_1),$ $ang(K_3,K_1,T_2))$	$ang(O,P,Target_2)$

7.2.2.1 Rule Transfer: Ringworld to 3 vs. 2 Keepaway

In this section we first detail how Rule Transfer between Ringworld and $25m \times 25m$ 3 vs. 2 Keepaway was performed. We then compare the results from using the three different advice utilization schemes and show that Extra Action is superior. Lastly, we detail a set of experiments showing that Rule Transfer, while it has multiple parameters, is not particularly sensitive to these parameters' settings.

Agents learn for 25,000 episodes in Ringworld and then record 20,000 (s,a) pairs, which take less than 1,000 episodes. After JRip learns a decision list, the rules are transformed via `Translate()` and the inter-task mappings (Table 7.4). Lastly, the decision list is used by Keepaway agents.

Table 7.5 shows one of the main results of this section: all three rule utilization methods can significantly increase all three transfer metrics. Furthermore, the asymptotic performance is not adversely affected by Rule Transfer for the best parameter settings when compared to learning without transfer. We conclude that the advice, provided as rules to the learner, can be successfully augmented over time to improve performance. Further, these results show that the Extra Action rule utilization method is slightly superior to the other two methods, and confirm that cross-domain transfer can be effective at increasing the speed of learning in Keepaway. Figure 7.8 shows learning in Keepaway without transfer and when using Extra Action Rule Transfer from Ringworld.

Note that in this study we ran all Keepaway experiments for 40 simulator hours. However, in informal experiments, it appeared that learning continued to improve

Table 7.5 A comparison of three rule utilization schemes to learning Keepaway without transfer. Each row is the average of 20 independent trials and shows the standard error (note that the top row uses the same settings as learning without transfer in Table 7.3 but with more trials). Numbers in **bold** are statistically better than learning without transfer at the 95% level, as determined via a Student's t-test.

Rule Transfer: Ringworld to 3 vs. 2 Keepaway

	Initial Performance	Asymptotic Performance	Accumulated Reward
		Without Transfer	
	7.8 ± 0.1	21.6 ± 0.8	756.7 ± 21.8
AddedConstant		Value Bonus	
5	$\mathbf{11.1 \pm 1.4}$	19.8 ± 0.6	722.3 ± 24.3
10	$\mathbf{11.5 \pm 1.7}$	$\mathbf{22.2 \pm 0.8}$	$\mathbf{813.7 \pm 23.6}$
InitialEpisodes		Extra Action	
100	$\mathbf{11.9 \pm 1.8}$	$\mathbf{23.0 \pm 0.5}$	$\mathbf{842.0 \pm 26.9}$
250	$\mathbf{11.8 \pm 1.9}$	$\mathbf{23.0 \pm 0.8}$	$\mathbf{827.4 \pm 33.0}$
Initial Episodes		Extra Variable	
100	$\mathbf{11.8 \pm 1.9}$	21.9 ± 0.9	$\mathbf{784.8 \pm 27.0}$
250	$\mathbf{11.7 \pm 1.8}$	$\mathbf{22.4 \pm 0.8}$	$\mathbf{793.5 \pm 22.2}$

policies after 40 hours (albeit slowly). Although the term "Final Performance" may be more appropriate, we use "Asymptotic Performance," and note that our metrics are only approximate. Given that many learning methods and function approximators are not guaranteed to converge, the Asymptotic Performance metric will often be an approximation.

As an example of the type of knowledge transferred from Ringworld to Keepaway, consider the transformed rule in Figure 7.7 which was observed in one trial. This rule demonstrates that the agent has learned that it should pass if a taker is close, there is not a taker very close to the target teammate, and the passing angle indicates that the teammate is open.

To further investigate the robustness of transfer from Ringworld to Keepaway, we perform a series of additional studies, as shown in Table 7.6, to determine the sensitivity of Rule Transfer to various parameter settings. First we try learning for

IF ((dist(K_1,T_1) <= 4) AND
(Min($dist(K_3,T_1), dist(K_3,T_2)$) >= 12.8) AND
($ang(K_3,K_1,T)$ >= 36))
THEN Pass to K_3

Fig. 7.7 An example transformed rule from Ringworld that would be difficult for a human to generate from domain knowledge alone

Table 7.6 This table shows Ringworld transfer with Extra Action rule usage after forcing the action advised by D_t for 100 episodes. The settings used previously (in Table 7.5) are shown in bold for comparison, each row is the average over 20 independent trials, and the standard error is shown.

Ringworld Sensitivity Analysis

Param	Initial Performance	Asymptotic Performance	Accumulated Reward
Learning Without Transfer			
	7.8 ± 0.1	21.6 ± 0.8	756.7 ± 21.8
Episodes of Ringworld Training before Recording Data			
20,000	10.1 ± 1.7	21.8 ± 1.3	762.5 ± 44.1
25,000	11.9 ± 1.8	23.0 ± 0.5	842.0 ± 26.9
30,000	12.0 ± 1.7	20.7 ± 5.0	793.9 ± 47.8
Ringworld's Ring Diameter (m)			
7.5	14.8 ± 2.4	20.0 ± 1.5	748.2 ± 53.6
8.5	13.5 ± 1.5	21.1 ± 1.2	776.8 ± 45.2
9.5	11.9 ± 1.8	23.0 ± 0.5	842.0 ± 26.9
10.5	9.4 ± 1.0	21.5 ± 1.3	757.7 ± 42.4
11.5	8.2 ± 1.3	20.1 ± 1.6	705.0 ± 41.6
Amount of recorded Ringworld Data			
5,000	12.2 ± 1.2	20.6 ± 4.9	765.1 ± 59.4
20,000	11.9 ± 1.8	23.0 ± 0.5	842.0 ± 26.9
40,000	11.2 ± 1.3	21.4 ± 1.3	776.4 ± 46.6
JRip Settings			
N=2, O=2	13.7 ± 1.7	20.9 ± 1.3	767.3 ± 44.3
N=100, O=2	10.7 ± 1.5	21.6 ± 1.2	784.3 ± 49.9
N=100, O=10	11.9 ± 1.8	23.0 ± 0.5	842.0 ± 26.9
N=2, O=10	14.0 ± 1.8	20.9 ± 1.3	763.3 ± 44.7

different amounts of time in Ringworld. When reducing the number of source task learning episodes to 20,000, the Ringworld agent has not yet plateaued; it is not surprising that the initial performance in Keepaway is slightly reduced (relative to learning for 25,000 episodes). Learning more after the Ringworld learner has plateaued (i.e., for 30,000 episodes) does not hurt performance. When using different ring diameters in Ringworld, the source task becomes less similar to 3 vs. 2 Keepaway, but all diameters do successfully improve one or more transfer metrics relative to learning without transfer.

We also performed experiments that examined the robustness of rule learning for transfer. In the first experiment we recorded different amounts of Ringworld data; less data would force more generalization while more data may cause overfitting. The last sensitivity analysis varied the parameters of JRip, again showing that the performance of the 4 metrics is not particularly sensitive to the rule learning settings, as they all outperform learning without transfer. The JRip parameter N is the

Table 7.7 This table describes the inter-task mapping used by `Translate()` to modify a decision list learned in the Knight Joust so that it can apply to Keepaway. It is very similar to the inter-task mapping introduced earlier in Section 5.3.4, Table 5.11, used to transfer from Knight Joust to 4 vs. 3 Keepaway with Value Function Transfer.

Inter-Task Mappings for Keepaway to Knight Joust

Keepaway	**Knight Joust**
χ_A	
Hold Ball	Forward
Pass to closest keeper	$Jump_{West}$
Pass to furthest keeper	$Jump_{East}$
χ_X	
$dist(K_1, T_1)$	$dist(P, O)$
$\text{Min}(ang(K_2, K_1, T_1)$ $ang(K_2, K_1, T_2))$	ang(West)
$\text{Min}(ang(K_3, K_1, T_1),$ $ang(K_3, K_1, T_2))$	ang(East)

minimum number of instances a rule must cover (JRip default = 2) and O is the number of optimization runs to increase generality (JRip default = 2). Thus, while there are a number of parameters tuned during Rule Transfer, the parameters proved easy to set in practice and were not critical to the method's success.

7.2.2.2 Rule Transfer: Knight Joust to Keepaway

In this section we present the results for transferring from Knight Joust to Keepaway using the inter-task mappings in Table 7.7.[4] Briefly, the intuition for these mappings is that the forward action is similar to the hold ball action because the player should take it whenever practical (i.e., executing the action does not soon end the episode). Note that the we have made West in the Knight Joust correspond to K_2 and East correspond to K_3, but either is reasonable, as long as the state variables and actions are consistent. When it is "too dangerous," the player instead jumps to the West or East, similar to passing the ball to K_2 or K_3. We first train the Knight Joust players for 50,000 episodes, as initial experiments showed that learners generally stopped learning after roughly this many episodes. The advice is utilized by Extra Action Rule Transfer in Keepaway (informal experiments showed that Value Bonus and Extra Variable under-performed Extra Action, as in Ringworld) and other parameters are unchanged from the previous section. The results from these experiments are presented in Table 7.8 and Figure 7.8.

[4] As mentioned in Section 4.5, the Knight Joust task in this set of experiments uses an older version of the task than was used for Value Function Transfer. In this task formulation, the rewards are +20 for taking the forward action, +20 for reaching the goal line, and +0 for any other action. In the Value Function Transfer experiments (Section 5.3.4) the rewards were +5, +50, and +0, respectively.

Table 7.8 Transferring from Knight Joust to Keepaway significantly improves the initial performance, but the other two metrics are not improved. All results are averaged over 20 independent trials and the standard error is shown. Numbers in **bold** are statistically different from learning without transfer at the 95% level, as determined via a Student's t-test.

Rule Transfer: Knight Joust to 3 vs. 2 Keepaway

Param	Initial Performance	Asymptotic Performance	Accumulated Reward
Without Transfer			
	7.8 ± 0.1	21.6 ± 0.8	756.7 ± 21.8
Extra Action			
100	$\mathbf{13.8 \pm 1.1}$	21.8 ± 1.2	758.5 ± 29.3
250	$\mathbf{13.5 \pm 0.9}$	21.6 ± 0.9	747.9 ± 25.3

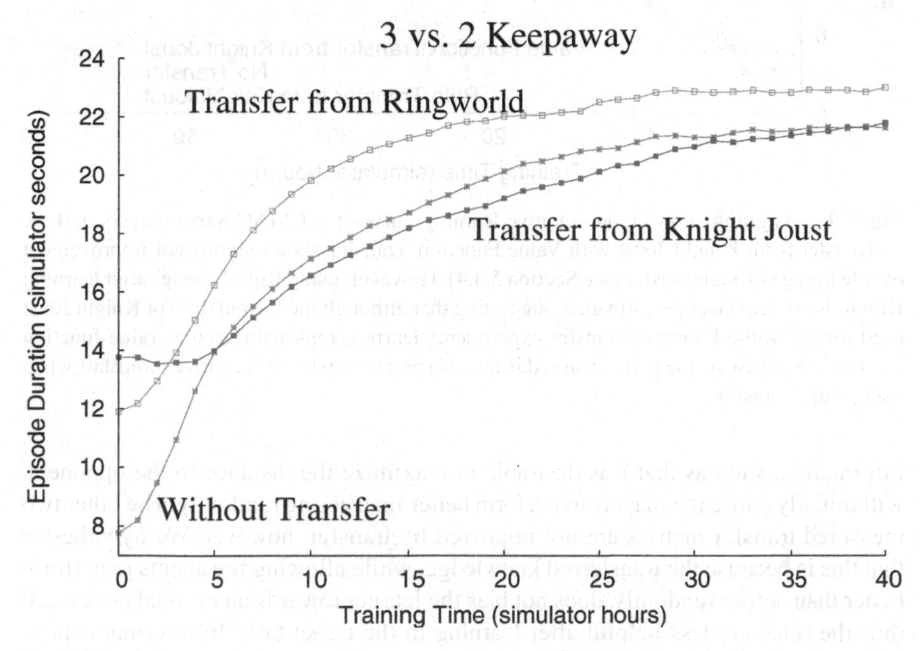

Fig. 7.8 Learning curves in 3 vs. 2 Keepaway, averaged over 20 trials, showing learning without transfer, learning with Extra Action from Ringworld after 100 episodes of following the rule-suggested actions, and learning with Extra Action from Knight Joust after 100 episodes of following the rule-suggested actions. These results were collected using version 0.6 of the Keepaway benchmark players.

The Knight Joust task is less similar to 3 vs. 2 Keepaway than Ringworld is to 3 vs. 2. There are many fewer state variables, a less similar transition function, and a very different reward structure. However, information from Knight Joust can significantly improve the initial performance of Keepaway players because very basic

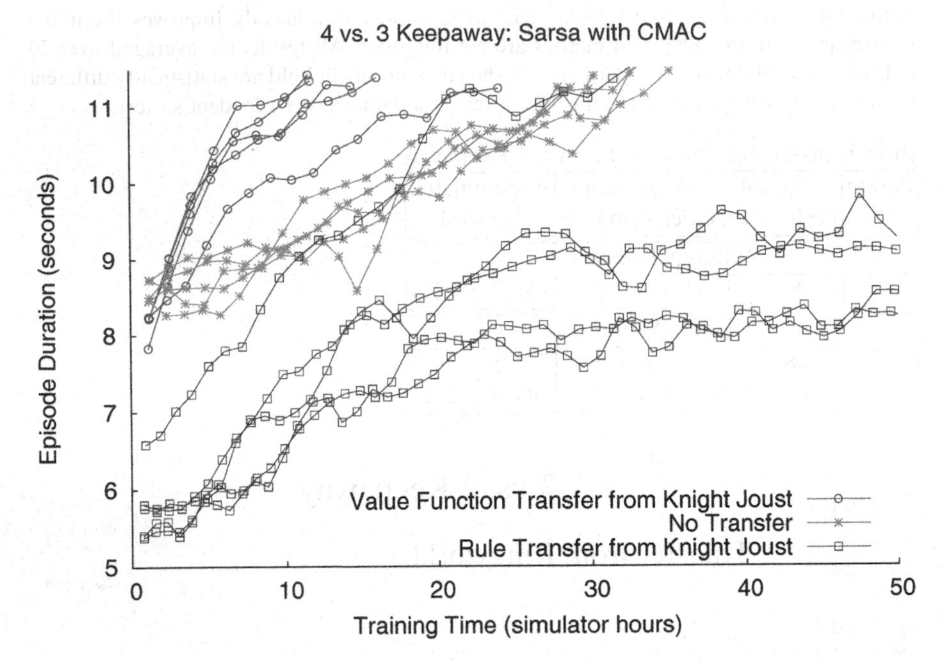

Fig. 7.9 This graph shows representative learning curves for CMAC Sarsa players in 4 vs. 3. Transfer from Knight Joust with Value Function Transfer shows significant improvement over learning without transfer (see Section 5.3.4). However, using Rule Transfer after learning Knight Joust decreases performance, suggesting that although the formulation of Knight Joust used for the Value Function Transfer experiments learns a reasonable action-value function for 4 vs. 3 Keepaway, the policy learned is not (as can be seen by the negative jumpstart when using Rule Transfer).

information, such as that it is desirable to maximize the distance to the opponent, will initially cause the players to perform better than acting randomly. The other two measured transfer metrics are not improved by transfer, however. We hypothesize that this is because the transferred knowledge, while allowing the agents to perform better than acting randomly, does not bias the learner towards an optimal policy and thus the rules are less helpful after learning in the target task. In informal experiments, after 40 hours of training in Keepaway, agents that had transferred advice from Ringworld were following the advice for 90% of the actions, while agents that transferred advice from Knight Joust were following the advice for only 85% of the actions, indicating that the Ringworld advice was more useful in Keepaway than Knight Joust advice.

To attempt to directly compare Value Function Transfer with Rule Transfer, we also replicated the experiment from Section 5.3.4. Knight Joust[5] is first learned for 50,000 episodes. Value Function Transfer uses ρ and the inter-task mappings

[5] To compare with Value Function Transfer, we use rewards of +5 for executing the forward action and +50 for reaching the goal line.

defined in Table 5.11 in Section 5.3.4 to initialize 4 vs. 3 Keepaway learners and then learns using Sarsa and CMAC function approximation. Rule Transfer uses JRip to learn rules summarizing the Knight Joust agent's policy, follow the rules for 100 episodes in 4 vs. 3, and then learn in 4 vs. 3 with the Extra Action utilization scheme. However, as shown in Figure 7.9, results using Rule Transfer were significantly worse than using Value Function Transfer or learning without transfer.

When visually examining the initial policies in 4 vs. 3, it appears that the transferred decision list consistently held the ball too long, causing it to be quickly lost to the takers. The resulting learning curves suggest that it takes significant experience to learn to compensate for this initial behavior, and is an example of negative transfer, which will be discussed further in the future work section of this monograph. These results suggest that this version of Knight Joust is unsuitable for Rule Transfer with a 4 vs. 3 target task, although the action-value function produced can be used to learn 4 vs. 3 Keepaway. To better compare Rule Transfer with Value Function transfer, the next section returns to transfer between 3 vs. 2 and 4 vs. 3 Keepaway.

7.2.3 Rule Transfer: 3 vs. 2 Keepaway to 4 vs. 3 Keepaway

In order to directly compare Rule Transfer with Value Function Transfer, we test both methods on 4 vs. 3 Keepaway using Sarsa learners with CMAC function approximation. Agents in 3 vs. 2 first train for 1,000 episodes. Value Function Transfer uses the saved action-value function in conjunction with ρ_{CMAC} to initialize 4 vs. 3 players. Rule Transfer records 10,000 actions[6] from the trained 3 vs. 2 players, forces the players to follow the transferred decision list for 100 episodes, and then learns in 4 vs. 3 using the Extra Action rule utilization method. Figure 7.10 shows the target task performance of all three sets of learners.

Rule Transfer has a significant jumpstart over both Value Function Transfer and No Transfer ($p < 0.05$). The total reward, calculated as the sum of the average performance at each hour, is 1016, 991, and 901 for Value Function Transfer, Rule Transfer, and No Transfer, respectively. While both transfer methods are improvements over not using transfer, Value Function Transfer provides a significant benefit to the total reward accumulated over Rule Transfer ($p < 0.05$). This result suggests that if both Value Function Transfer and Rule Transfer are applicable, Value Function Transfer should be preferred (unless the initial performance in the target task was critical).

7.2.4 Rule Transfer Summary

This section of the monograph has introduced Rule Transfer along with three different advice utilization methods. We conducted experiments to demonstrate that Rule Transfer is effective for cross-domain transfer, showing that both Ringworld and Knight Joust could improve learning in 3 vs. 2 Keepaway. The Ringworld task

[6] 10,000 actions take roughly 1 simulator hour of source task time, or 300 episodes.

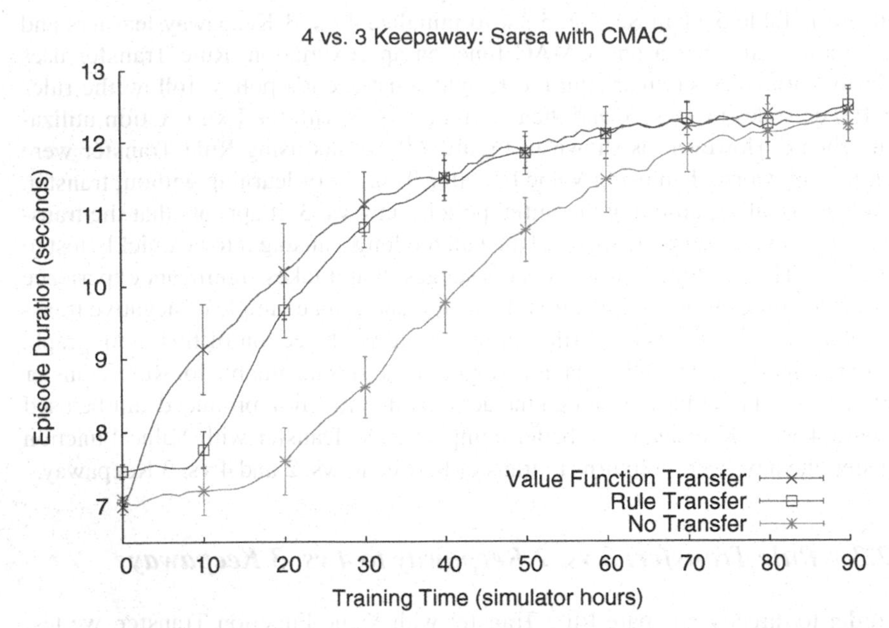

Fig. 7.10 This graph shows the average performance of CMAC Sarsa players in 4 vs. 3. Both transfer methods are significant improvements over No Transfer. Rule Transfer has a higher jumpstart than Value Function Transfer, but Value Function Transfer has a higher total reward. Each curve is averaged over 15 independent trials and bars show the standard error. These results were collected using version 0.6 of the Keepaway benchmark players.

was constructed directly from data gathered in the target task while the Knight Joust task was chosen as intuitively related to Keepaway. Additionally, we compare Rule Transfer to Value Function Transfer in Keepaway, showing that although Rule Transfer can have higher initial performance in the target task, Value Function Transfer yields higher total reward.

The cross-domain transfer experiments in this section begin to demonstrate the flexibility of Rule Transfer; unlike TL methods presented in previous chapters, agents were able to transfer knowledge successfully irrespective of the underlying function approximator's representation. This flexibility is particularly noticeable when learning Ringworld and Knight Joust, as they are simple tasks which can be learned very quickly with tabular function approximator. Lastly, because rules are learned that summarize the source task policy, the relative magnitude of the rewards in the source and target task do not affect transfer, potentially allowing for more flexibility when selecting pairs of tasks to transfer between.

In the next section we discuss Representation Transfer, a method designed to be even more flexible than Rule Transfer, in terms of learning methods, function approximators, and scenarios where it is applicable.

7.3 Representation Transfer

A key component of any RL algorithm is the underlying *representation* used by the agent for learning (e.g., its function approximator or learning algorithm). TL approaches, including those presented thus far in this monograph, generally assume that the agent will use a similar (or even the same) representation to learn the target task as it used to learn the source. However, this assumption may not be necessary or desirable. This section considers a different question: is it possible, and desirable, for agents to use different representations in the target and source? The TL methods presented thus far can successfully transfer knowledge between different tasks, while this section defines and provides algorithms for this novel problem of *representation transfer* and contrasts it with the more typical task transfer.

The motivation for transferring knowledge between tasks is clear: it may enable quicker and/or better learning on the target task after having learned on the source. Our motivations for representation transfer are similar, though perhaps a bit more subtle.

The first motivation for equipping an agent with the flexibility to learn with different representations is procedural. Suppose an agent has already trained on a source task with a certain learning method and function approximator, but the performance is poor. A different representation could allow the agent to achieve higher performance. If experience is expensive (e.g., wear on the robot, data collection time, or cost of poor decisions), it is preferable to leverage the agent's existing knowledge to improve learning with the new representation and minimize sample complexity.

A second motivating factor is learning speed: changing representations partway through learning may allow agents to achieve better performance in less time. SOAR [Laird et al. (1987)] can use multiple descriptions of planning problems and search problems, generated by a human user, for just this reason. We will show in this section that it is advantageous to change internal representation while learning in some RL tasks, as opposed to using a fixed representation, to achieve higher performance more quickly.

The final motivation for representation transfer is human psychological experiments. Agents' representations are typically fixed when prototyped, but studies show (c.f., [Simon (1975)]) that humans may change their representation of a problem as they gain more experience in a particular domain. While our system does not allow for automatic generation of a learned representation, this method addresses the necessary first step of being able to transfer knowledge between two representations.

Using multiple representations to solve a problem is not a new idea. For instance, Kaplan's production system [Kaplan (1989)] was able to simulate the representation shift that humans often undergo when solving the *mutilated checkerboard* [McCarthy (1964)] problem. Other work [Fink (1999)] used libraries of problem solving and "problem description improvement" algorithms to automatically change representations in planning problems. *Implicit imitation* [Price and Boutilier (2003)] allows an RL agent to train while watching a mentor with similar actions, but this method does not directly address internal representation differences. Additionally, all training is done on-line; agents using imitation do not initially perform better than learning without transfer. Our method of training offline from saved experience is more similar to the idea of *replayed TD* [Mahadevan and Connell (1991)], a method to improve the rate of learning by reusing experience in a single agent.

None of these methods directly address the problem of transferring knowledge between different representations in an RL setting. By using Representation Transfer, different representations can be leveraged so that better performance can be more quickly learned, possibly in conjunction with other RL speedup methods.

This section's main contributions are to introduce representation transfer, to provide a set of algorithms for tackling representation transfer problems, and to empirically demonstrate the efficacy of these algorithms in Keepaway. In order to test Representation Transfer, we train on the same tasks with different learning algorithms, functions approximators, and parameterizations of these function approximators, and then demonstrate that transferring the learned knowledge among the representations is both possible and beneficial. Lastly, we show that the algorithms can be used for successful task transfer, underscoring the relatedness of representation transfer and task transfer.

This section presents two algorithms for addressing representation transfer problems, where the source and target representations differ. We define an agent's *representation* as the learning method used, the function approximator used, and the function approximator's parameterization.[7] As an example, suppose a source agent uses Q-Learning with a neural network function approximator that has 20 hidden nodes. The first algorithm, *Complexification*, is used to:

1. Transfer between different parameterizations (e.g., change to 30 hidden nodes)

The second, *Offline Representation Transfer*, may be used for:

2. Transfer between different function approximators (e.g., change to a radial basis function approximator)
3. Transfer between different learning methods (e.g., change to direct policy search learning)
4. Transfer between tasks with different actions and state variables (i.e., task transfer)

We refer to scenarios 1 and 2 as *intra-policy-class transfer* because the policy representation remains constant. Scenario 3 is a type of *inter-policy-class transfer*, and Scenario 4 is task transfer, as discussed in previously introduced TL methods.

The Representation Transfer method encompasses a number of different algorithms, which will be detailed later. However, it is important to remember that all are similar in that they use knowledge learned in a source representation to improve learning in a target representation, and that they may be combined with inter-task mappings.

[7] It is more common to consider an agent's representation as the state variables it observes when interacting with the world and possibly the actions it can execute. In this monograph we expand the definition to point out that the learning method, the function approximator, and the function approximator's parameterization all affect what knowledge is learned or how it is stored, and is thus a candidate for transfer.

Algorithm 11. Complexification

1: Train with a source representation and save the learned FA_{source}
2: **while** target agent trains on a task with FA_{target} **do**
3: **if** $Q(s,a)$ needs to use at least one uninitialized weight in FA_{target} **then**
4: Find the set of weights W that would be used to calculate $Q(s,a)$ with FA_{source}
5: Set any remaining uninitialized weight(s) in FA_{target} needed to calculate $Q(s,a)$ to the average of W

7.3.1 Complexification

Complexification is a type of Representation Transfer where the function approximator is changed over time to allow for more representational power. Consider, for instance, the decision of whether to represent state variables conjunctively or independently. A linear interpolation of different state variables may be faster to learn, but a conjunctive representation has more descriptive power. Using Complexification, the agent can learn with a simple representation initially and then switch to a more complex representation later. Thus the agent can reap the benefits of fast initial training without suffering decreased asymptotic performance.

Algorithm 11 describes the process for transferring between value function representations with different parameterizations of state variables, such as function approximators with different dimensionalities. We abbreviate function approximator as *FA* for readability. The weights (parameters) of a learned FA are used as needed when the agent learns a target value function representation. If the target representation must calculate $Q(s,a)$ using a weight which is set to the default value rather than a learned one, the agent uses the source representation to set the weight. Using this process, a single weight from the source representation can be used to set multiple weights in the target representation.

Complexification is similar to Value Function Transfer in that it directly transfers weights from a source to a target and is similar to Q-Value Reuse because it performs this transfer on-line, rather than as a step between learning the source and target. The idea of Complexification is also related to *Incremental Feature-Set Augmentation* (IFSA) [Ahmadi et al. (2007)], which is designed to add state variables over time, rather than to transfer between different function approximator parameterizations.

Note that this algorithm makes the most sense when used for function approximators that exhibit *locality*: line 5 in Algorithm 11 would execute once and initialize all weights when using a fully connected neural network. Thus we employ Algorithm 11 when using a function approximator which has many weights but in which only a subset are used to calculate each $Q(s,a)$ (such as in a CMAC). Also, note that line 5 is similar to the averaging step in the ρ_{CMAC} for Value Function Transfer in Section 5.3.2.

We will utilize Complexification on a task which requires a conjunctive representation for optimal performance. This provides an existence proof that Complexification can be effective at reducing both the target representation training time and the total training time.

7.3.2 *Offline Representation Transfer*

The key insight behind *Offline Representation Transfer* (ORT) is that an agent using a source representation can record some information about its experience using the learned policy as was done in TIMBREL and Rule Transfer. The agent may record s, the perceived state; a, the action taken; r, the immediate reward; and/or $Q(s,a)$, the long-term expected return. Then the agent can learn to mimic this behavior in the target representation through off-line training (i.e., without more interactions with the environment). The agent is then able to learn better performance faster than if it had learned the target representation without transfer. We consider three distinct scenarios where ORT algorithms could be utilized:

1. Intra-policy-class Representation Transfer (Algorithm 12): The representation differs by function approximator.
2. Inter-policy-class Representation Transfer (Algorithms 13 & 14): The representation changes from a value function learner to a policy search learner, or vice versa.
3. Task transfer (Algorithm 15): The representation remains constant but the tasks differ.

Note that this is not an exhaustive list; it contains only the variants which we have implemented and tested. (For instance, intra-policy-class Representation Transfer for policy learners is similar to Algorithm 12, and task transfer combined with inter-policy-class transfer is likewise a straightforward extension of the ORT method.) The ORT algorithms presented are necessarily dependant on the details of the representation used. Thus ORT may be appropriately thought of as a meta-algorithm and we will show later how it may be instantiated for specific learning methods and specific function approximators.

Algorithm 12 describes intra-policy-class transfer for value function methods with different function approximators. The agent saves $\langle s,a,Q(s,a)\rangle$ tuples and then trains offline with the target representation to predict those saved Q-values, given the corresponding state and action. Here offline training still utilizes a TD update, but the target Q-values are set by the recorded experience. When considering inter-policy-class transfer between a value function and a policy search method, the primary challenge to overcome is that the learned function approximators represent different concepts: a value function by definition contains more information because it represents not only the best action, but also its expected value.

Inter-policy-class transfer from a policy to a value function (Algorithm 13) works by recording n (s, a, r) and then training a TD learner offline by (in effect) replaying the learned agent's experience, similar to Algorithm 12. Line 4 uses the history to calculate q_i. In the undiscounted episodic case, the optimal predicted return from time t_i, q_i, is $\sum_{t_i<t\leq t_{EpisodeEnd}} r_t$, and can thus be found by summing recorded rewards until the end of the episode is reached (i.e., the Monte Carlo return). Similarly, the

Algorithm 12. Offline Representation Transfer: Value Functions

1: Train with a source representation
2: Record n $(s, a, q(s,a))$ tuples while the agent acts
3: **for all** n tuples **do**
4: Train offline with target representation, learning to predict $Q_{target}(s_i, a_i) = q(s_i, a_i)$ for all $a \in A$
5: Train on-line using the target representation

Algorithm 13. Offline Representation Transfer: Policies to Value Functions

1: Train with a source representation
2: Record n (s, a, r) tuples while the agent acts
3: **for all** n tuples **do**
4: Use history to calculate the return, q_i, from s_i
5: Train offline with target representation, learning to predict $Q_{target}(s_i, a_i) = q_i$
6: **for all** $a_j \in A$ s.t. $a_j \neq a_i$ **do**
7: **if** $Q_{target}(a_j) > Q_{target}(a_i)$ **then**
8: Train to predict $Q_{target}(a_j) = c \times q_i$
9: Train on-line using the target representation

Algorithm 14. Offline Representation Transfer: Value Functions to Policies

1: Train with a source representation
2: Record n (s, a) tuples while the agent acts
3: **for all** n tuples **do**
4: Train offline with target representation, learning $\pi_{target}(s_i) = a_i$
5: Train on-line using the target representation

discounted non-episodic case would sum rewards, multiplied by a discount factor. Line 8 is used to generate some initial Q-values for actions not taken. If an action was not taken, we know that its Q-value was lower, but cannot know its exact value since the source policy learner does not estimate Q-values. A parameter c, in the range of $(0,1)$, determines how much to penalize actions not selected.

Inter-policy-class transfer between a value function and a policy search learner (Algorithm 14) first records n (s, a) tuples and then trains a direct policy search learner offline so that π_{target} can behave similarly to the source learner. Here, offline training simply means using the base learning algorithm to learn a policy that will take the same action from a given state as was taken in the saved experience.

Lastly, we present an algorithm for inter-task transfer using a value function approximator (Algorithm 15). Specifically, we assume that we have a pair of tasks that have different action and state variable spaces, but are related by two inter-task mappings. This algorithm is similar to Q-Value Reuse, but it uses saved data to train a new function approximator offline rather than directly reusing an old FA.

Algorithm 15. Offline Representation Transfer: Task Transfer for Value Functions

1: Train on a source task
2: Record n $(s, a, q(s,a))$ tuples while the agent acts
3: **for all** n tuples **do**
4: Construct s_{target} such that for each state variable $x_i \in s_{target}$, x_i has the value of state variable $\chi_X(x_i)$ in s
5: Train offline in target task, learning to predict $Q_{target}(s_{target}, a_{target}) = q(s,a)$ for all a_{target} where $\chi_A(a) = a_{target}$
6: Train on-line using the target task

7.3.3 *Overview of Representation Transfer Results*

This section presents empirical results showing that Complexification and the four variations of ORT can successfully transfer knowledge. Representation Transfer encompasses a number of algorithms, all of which are fundamentally similar because they can be combined with inter-task mappings and are designed to effective re-use knowledge, there are many possible tests that could be performed. We select a set of representative experiments to test each Representation Transfer method (Algorithms 11–15) on Keepaway tasks using RL algorithms and function approximators from previous chapters. While this list of experiments does not exhaust the scenarios where Representation Transfer could be used, these experiments provide a set of examples that support the claim that Representation Transfer can be effectively used to improve learning. Specifically, we test:

1. Complexification in XOR Keepaway with a Sarsa learner utilizing a CMAC function approximator
2. ORT for Value Functions in 3 vs. 2 Keepaway between RBF and neural network Sarsa learners
3. ORT for Value Functions in 3 vs. 2 Keepaway between neural network and RBF Sarsa learners
4. ORT for Policies to Value Functions in 3 vs. 2 Keepaway between NEAT and Sarsa learners
5. ORT for Value Functions to Policies in 3 vs. 2 Keepaway between Sarsa and NEAT learners
6. ORT for Task Transfer with Value Functions between 3 vs. 2 and 4 vs. 3 Keepaway

The experiments that test these six scenarios are summarized in Table 7.9.

We consider two related goals for both representation and task transfer problems. We first show that all of the representation transfer methods presented can reduce the training time in the target task. Additionally, experiments 1, 5, and 6 show that the total training time may also be reduced, a significantly more difficult transfer goal. For representation transfer, that means that an agent can improve performance on a single task by switching internal representations partway through learning, rather than using a single representation for an equivalent amount of time.

Table 7.9 This table summarizes the Representation Transfer experiments in this section of the monograph. The final experiment, marked with $*$, uses Offline Representation Transfer to transfer between 3 vs. 2 Keepaway and 4 vs. 3 Keepaway.

Method Name: Representation Transfer: Complexification				
Scenario: Complexification is applicable when the parameterization of a function approximator can change over time and the function approximator exhibits locality.				
Task	**Source Representation**	**Target Representation**	**Algorithm Tested**	**Section**
XOR Keepaway	Sarsa, CMAC, individually tiled	Sarsa, CMAC conjunctively tiled	11	7.3.4

Method Name: Representation Transfer: Offline Representation Transfer				
Scenario: Offline Representation Transfer is applicable when the representation changes in a task, or for inter-task transfer. Offline Representation Transfer is composed of a set of very flexible methods to enable transfer between a number of different learning representations.				
Task	**Source Representation**	**Target Representation**	**Algorithm Tested**	**Section**
3 vs. 2 Keepaway	Sarsa, ANN	Sarsa, RBF	12	7.3.5.1
3 vs. 2 Keepaway	Sarsa, RBF	Sarsa, ANN	12	7.3.5.1
3 vs. 2 Keepaway	NEAT, ANN	Sarsa, RBF	13	7.3.5.2
3 vs. 2 Keepaway	Sarsa, RBF	NEAT, ANN	14	7.3.5.2
Task Transfer*	Sarsa, RBF	Sarsa, RBF	15	7.3.6

All learning curves presented in this section each average ten independent trials. The x-axis shows the number of Soccer Server simulator hours. The y-axis shows the average performance of the keepers by showing the average episode length in simulator seconds. Error bars show one standard deviation. (Note that we only show error bars on alternating curves for readability.) All parameters chosen in this section were selected via experimentation with a small set of initial test experiments.

7.3.4 Complexification in XOR Keepaway

This section demonstrates that Complexification (Algorithm 11) can improve performance on a Keepaway task, relative to learning with a single representation. XOR

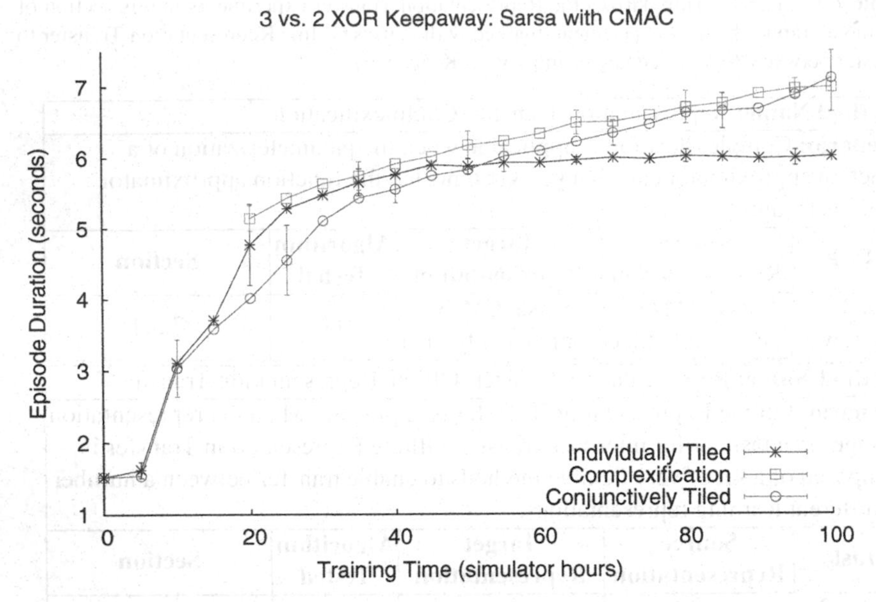

Fig. 7.11 Learning with Complexification outperforms learning with individually tiled CMACs without transfer and partially conjunctive CMACs without transfer. Complexification is used to transition from an individually tiled representation to a conjunctively tiled representation after learning for 20 simulator hours.

Keepaway was designed so that the agent's internal representation must be capable of learning an "exclusive or" to achieve top performance (see Section 4.3.2), and is thus a prime candidate for Complexification.[8]

To master the XOR Keepaway task we use Sarsa to learn with CMAC function approximators, with both independent and conjunctively tiled parameterizations. The independently tiled players use 13 separate CMACs, one for each state variable. The conjunctively tiled players use 10 separate CMACs, 9 of which independently tile state variables. The tenth CMAC is a conjunctive tiling of the remaining 4 state variables: $d(K_1,T_1)$, $d(K_2,T)$, $ang(K_2)$, and $d(K_1,K_2)$. We train the independently tiled players for 20 simulator hours and then save the weights in their CMAC function approximators. To get a small performance improvement, we set all zero weights to the average weight value, a method previously shown (see Section 5.3.2) to improve CMAC performance. We then train the conjunctively tiled CMAC players, using the previously learned weights as needed, as per Algorithm 11.

Agents learn best when the four relevant state variables are conjunctively tiled: Figure 7.11 shows that players learning with conjunctive function approximators outperform the players using independently tiled function approximators. However,

[8] In informal experiments, Complexification did not improve the benchmark 3 vs. 2 Keepaway task performance, likely because it can be learned well when all state variables are considered independently [Stone et al. (2005)].

agents using independently tiled state variables are initially able to learn faster. Agents trained with independent CMACs for 20 hours can transfer to a conjunctive representation via Algorithm 11, significantly outperforming players that only use the independent representation. A Student's t-test confirms that this performance increase is statistically different from learning without transfer with independently tiled CMACs after 40 hours of training.

Additionally, the total training time required is decreased by Complexification relative to learning only the conjunctive tiling. Even when source agent training time is also taken into account, Complexification significantly outperforms learning without transfer with the conjunctive representation until 55 hours of training time has elapsed. Examined differently, learning without transfer with a conjunctive tiling takes 55 hours to reach a 6.0 second average episode length, while an agent using both source and target representations take a total of only 45 hours, an 18% reduction in learning time.

This result shows that in the XOR Keepaway task, using Complexification to transfer knowledge between two different representations can outperform using either representation alone for the equivalent amount of training time.

7.3.5 *Offline Representation Transfer in 3 vs. 2 Keepaway*

When learning 3 vs. 2 Keepaway, ORT algorithms may be used to transfer knowledge between different function approximators and between different policy representations. We first show that if a source representation has been learned and a target representation utilizes a different function approximator, ORT can successfully reduce the target's training time. We next demonstrate that the source and target may differ both by function approximator and by type of learning method and still utilize ORT. In all experiments, the target's training is improved. One experiment also shows transfer benefit in both learning scenarios: both the target training and the total training are improved.

7.3.5.1 Intra-Policy Offline Representation Transfer

This section uses Algorithm 12 to transfer between Sarsa players with different function approximators, allowing the agents to effectively reuse knowledge. To demonstrate intra-policy transfer, we first train Sarsa players using a neural network on 3 vs. 2 Keepaway for 20 simulator hours and then record 20,000 (s, a) tuples, which took roughly 1.0 simulator hours. TD-RBF players are then trained offline by iterating over all tuples 5 times and updating $Q(s_i,a)$, where $a \in A$, via Sarsa with a learning rate of 0.001 (set after trying 4 different common parameter values). The agent is then able to learn in the new representation by replaying data gathered when training with the old representation. Using Algorithm 12, this process takes roughly 8 minutes of wall clock time. Figure 7.12 shows that representation transfer from neural network players outperforms RBF players learning without transfer. Differences graphed are statistically significant for times less than 11 simulator hours.

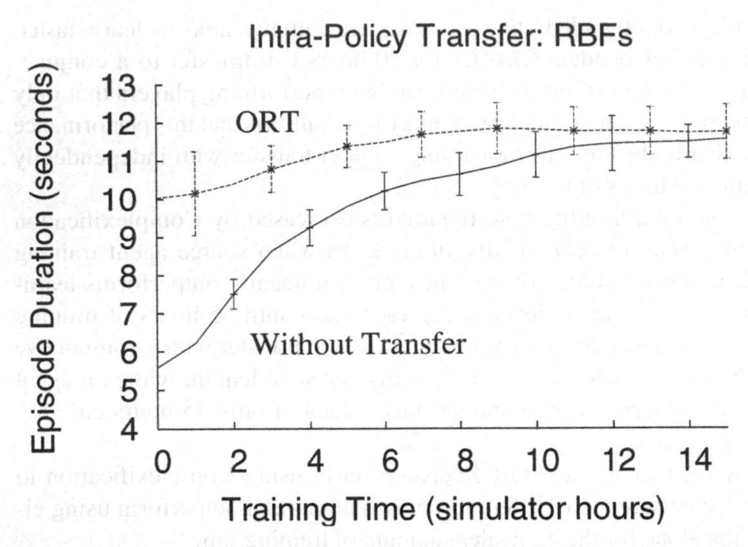

Fig. 7.12 RBF players utilize ORT from neural networks to outperform RBF players without transfer

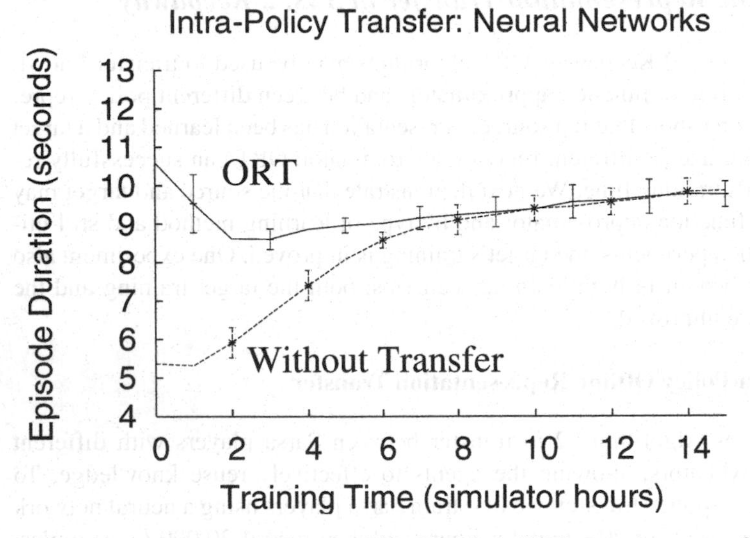

Fig. 7.13 Neural network players utilize ORT from RBF players to outperform neural network players without transfer

The complimentary experiment trains RBF players for 20 simulator hours and then saves 20,000 tuples. We then train the neural network players offline by iterating over all tuples five times. We found that updating $Q(s,a_i)$ for $a_i \neq a$ was not as efficient as updating only the Q-value for the action selected in a state. This difference is likely due to the non-locality effect of neural networks — changing a single weight may affect all output values. Figure 7.13 shows how representation transfer helps improve the performance of the neural network players. The differences are statistically significant for times less than 8 simulator hours and the offline representation transfer training took less than 1 minute of wall clock time.

7.3.5.2 Inter-Policy Offline Representation Transfer

This section uses Algorithms 13 and 14 to transfer between TD and policy search learners. One result will show that past knowledge can be effectively reused (the target task time scenario), and the second result will show that learning with two representations outperforms learning with either representation alone (representing success in both the target task time and total time scenarios).

To demonstrate transfer from policy to an action-value function, we first train NEAT keepers for 500 simulator hours in the 3 vs. 2 Keepaway task and then use representation transfer to initialize RBF players via offline Sarsa training. We found that the value-function agents needed to learn a more complex representation and used 50,000 tuples (which takes roughly 2.6 simulator hours to record). If $Q(s_i,a') > Q(s_i,a_i)$, where a' was an action not chosen by the source agent, we set a target value of $Q(s_i,a') = 0.9 \times Q(s_i,a_i)$.[9] The offline training, as described previously, takes roughly 4 minutes of wall clock time.

Figure 7.14 shows that the RBF players using representation transfer from learned NEAT representations initially have much higher performance. Training causes an initial drop in performance as the Q-values, and therefore the current policy, are changed to more accurately describe the task. However, performance of the players using representation transfer is statistically better than those learning without transfer, until 7 simulator traing hours. After 7 simulator hours, the performance difference between using representation transfer and learning without transfer is not significant. This result shows that if one has trained policies, it is advantageous to use them to initialize TD agents, particularly if the training time in the target representation is short or if the on-line reward is critical.

The reverse experiment trains 3 vs. 2 Keepaway using the value function RBF players for 20 simulator hours. After learning, one of the keepers saves 1,000 tuples, and we use inter-policy representation transfer to initialize a population of 100 policies offline. NEAT trains offline for 100 generations with a fitness function that sums the number of times the action predicted by NEAT from a given state matches that action that had been recorded. After the target representation keepers have finished learning, we evaluate the champion from each generation for 1,000 episodes to graph the learned policy performances. Figure 7.15 shows that NEAT players utilizing representation transfer outperform NEAT players learning without transfer. This result is particularly dramatic because TD-RBF players initially train much faster than NEAT players. The 20 hours of simulator time spent training the RBF players and the roughly 0.1 simulator hours to collect the 1,000 tuples are not reflected in this graph.

[9] Recall that the only information we have regarding the value of non-chosen actions are that they should be lower valued than selected actions. However, setting those values too low may disrupt the function approximator so that it does not generalize well to unseen states. $c = 0.9$ was chosen after initial tests were run on three different parameter values.

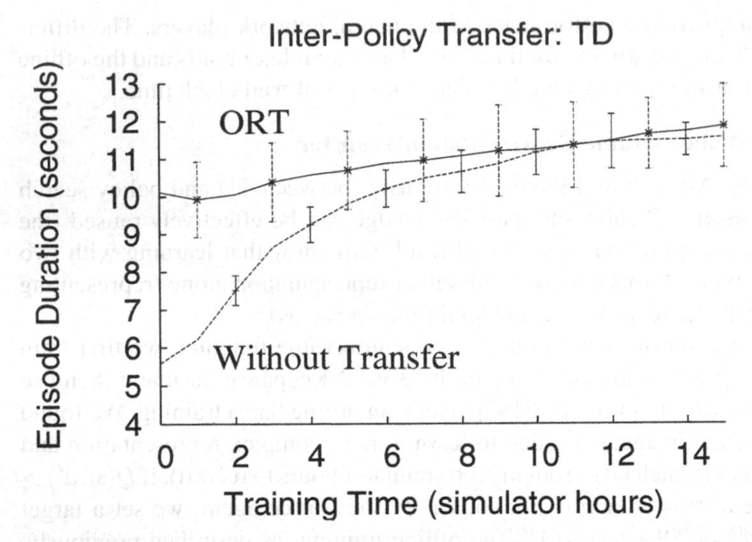

Fig. 7.14 RBF players using ORT from NEAT players outperform RBF players without transfer

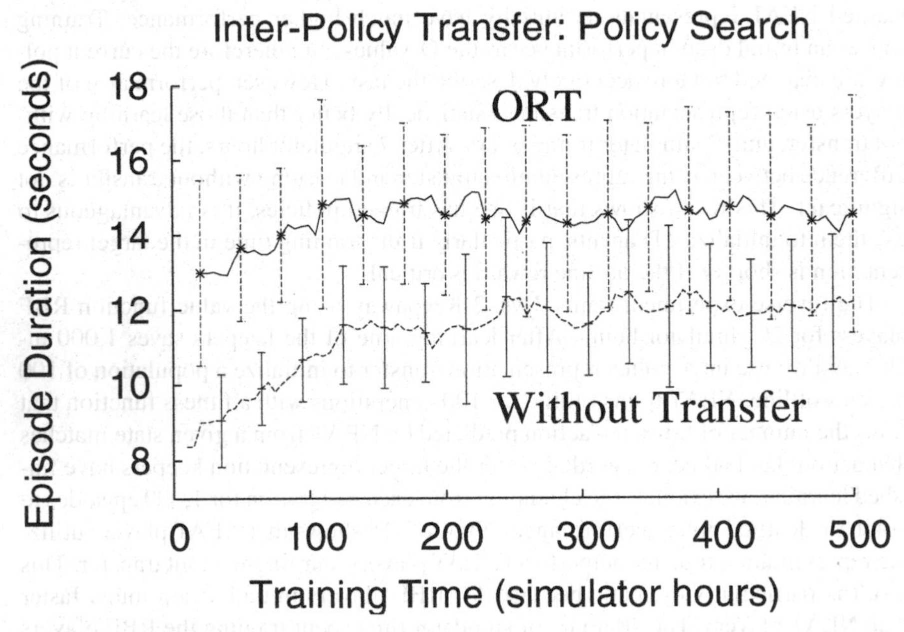

Fig. 7.15 ORT can initialize NEAT players from RBF players to significantly outperforms learning without transfer

The difference between learning with and without representation transfer is statistically significant for all points graphed in Figure 7.15 (except for 490 simulator hours) and the total training time needed to reach a pre-determined performance threshold in the target task has been successfully reduced. For instance, if the goal is to train a set of agents to possess the ball in 3 vs. 2 Keepaway for 14.0 seconds via NEAT, it takes approximately 700 simulator hours to learn without transfer (not shown). The total simulator time needed to reach the same threshold using ORT is less than 100 simulator hours. Additionally, the best learned average performance of 15.0 seconds is better than the best performance achieved by NEAT learning without transfer in 1000 simulator hours [Taylor et al. (2006)].

These experiments focus on sample complexity under the assumption that agents operating in a physical world are most affected by slow sample gathering. If computational complexity were taken into account, representation transfer would still show significant improvement. Although we did not optimize for it, the wall clock time for representation transfer's offline training when transferring from TD-RBF players to NEAT players was only 4.3 hours per trial. Therefore, representation transfer would still successfully improve performance if our goal had been to minimize wall clock time.

7.3.6 *Offline Representation Transfer for Task Transfer*

In this section, an experiment demonstrates that ORT is able to meet both transfer scenario goals (reducing the target task learning time and the total learning time) when transferring between 3 vs. 2 and 4 vs. 3 Keepaway, successfully performing task transfer. This result suggests both that ORT is a general algorithm that may be applied to both representation transfer and task transfer and that other representation transfer algorithms may work for both types of transfer.

To transfer between 3 vs. 2 and 4 vs. 3, we use χ_X and χ_A as defined previously (Section 5.1.1, Table 5.1). 3 vs. 2 players learning with Sarsa and RBF function approximators are trained for 5 simulator hours. The final 20,000 tuples are saved at the end of training (taking roughly 2 simulator hours). 4 vs. 3 players, also using Sarsa and RBF function approximators, are initialized by training offline using Algorithm 15, where the inter-task mappings are used to transform the experience from 3 vs. 2 so that the states and actions are applicable in 4 vs. 3. The batch training over all tuples is repeated 5 times.

Figure 7.16 shows that ORT reduces the target task training time, meeting the goal of transfer in the first scenario. The performance of the learners using ORT is better than that of learning without transfer until a time of 31 simulator hours. Furthermore, the total time is reduced when accounting for the 5 hours of training in 3 vs. 2. In this case, the ORT agents statistically outperform agents training without transfer during hours 10 – 25. Put another way, it will take agents learning without transfer an average of 26 simulator hours to reach a hold time of 7.0 seconds, but agents using ORT will use a total time of only 17 simulator hours to reach the same performance level.

When the Complexification algorithm is used for task transfer between 3 vs. 2 and 4 vs. 3, it can make use of ρ_X and ρ_A similar to how Value Function Transfer uses the inter-task mappings. The main difference between this approach and Value Function

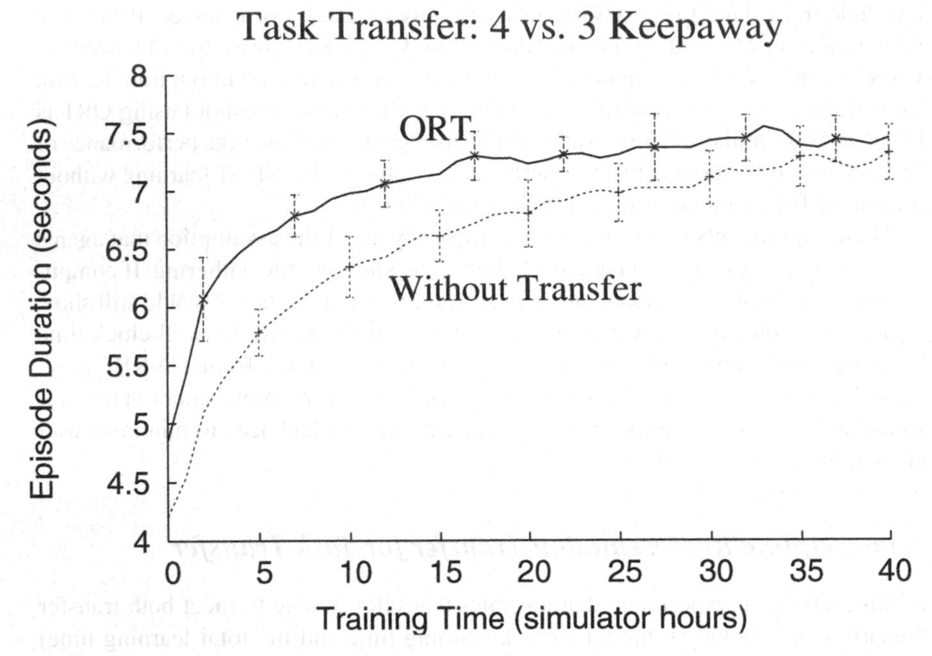

Fig. 7.16 ORT successfully reduces training time for task transfer between 3 vs. 2 and 4 vs. 3 Keepaway

Transfer is that Complexification transfers the weights on-line while the agent interacts with the target task, whereas Value Function Transfer copies weights *after* learning the source but *before* learning the target task. Thus experiments demonstrating Complexification for task transfer are unnecessary.

7.3.7 Representation Transfer Summary

This section has formulated the problem of representation transfer, a related, but distinct, problem from task transfer. We have presented two algorithms to transfer knowledge between internal representations. Five different experiments give positive examples of situations where representation transfer can improve agent performance relative to learning without transfer. Two of these experiments show that transfer can significantly reduce the total training time as well; in these two cases, it is better to learn with two representations in series than either of the representations individually. Additionally, we demonstrate that representation transfer algorithms can be used to reduce both target and total training times for task transfer. Experiments were conducted in 3 vs. 2 Keepaway, XOR Keepaway, and 4 vs. 3 Keepaway, using Sarsa and NEAT as representative learning algorithms and CMAC, RBFs, and neural networks as representative function approximators. Of the transfer algorithms presented in this monograph, the representation transfer methods are by far the most flexible in terms of source and target task agent requirements.

Table 7.10 This table summarizes the six TL methods introduced in this monograph, along with references to the relevant sections

Method: Value Function Transfer, Algorithms 6–8, Sections 5.2–5.4
Scenario: Value Function Transfer is only applicable when both the source task agent and target task agent use TD learning and both use the same type of function approximation. Results suggest that learning speed is improved more more effectively than by other TL methods.
Tasks: 3 vs. 2 Keepaway, ..., 7 vs. 6 Keepaway, 3 vs. 2 Flat Reward, 3 vs. 2 Giveaway, and Knight Joust **Learners:** Sarsa **Function Approximators:** Tabular, CMAC, RBF, and ANN
Method: Q-Value Reuse, Equation 5.1, Section 6.1
Scenario: This method is applicable only when all agents use TD learning. Agents are not required to use the same function approximator. This method is not as effective as Value Function Transfer and may require extra memory due to requiring multiple function approximators, but it is more flexible because agents may use different function approximators.
Tasks: Standard 2D Mountain Car, Standard 3D Mountain Car, 3 vs. 2 Keepaway, and 4 vs. 3 Keepaway **Learners:** Sarsa **Function Approximators:** CMAC
Method: Policy Transfer, Algorithm 9, Section 6.2
Scenario: Policy Transfer is applicable when the source and target task agents use direct policy search with ANN action selectors.
Tasks: 2-job-type SJS, 4-job-type SJS, 3 vs. 2 Keepaway, and 4 vs. 3 Keepaway **Learners:** NEAT **Function Approximators:** ANN
Method: TIMBREL, Algorithm 10, Section 7.1
Scenario: TIMBREL can be used when the target task agent uses instance-based RL. The source task agent may use any RL method. A sufficient number of instances must be recorded in the source task to approximate a model, which may require a significant amount of memory. This method uses source task instances to learn a target task model as the agent acts, likely increasing the computational requirements of the target task agent.
Tasks: Low Power 2D Mountain Car, No Goal 2D Mountain Car, High Power 2D Mountain Car, Low Power 3D Mountain Car, 3 vs. 2 Keepaway, and 4 vs. 3 Keepaway **Learners:** Fitted R-MAX **Function Approximators:** Instance based

Table 7.10 *(continued)*

Method: Rule Transfer, Section 7.2
Scenario: Rule Transfer can be used whenever the target task agent uses value function learning. Rules allow agents to transfer between different RL algorithms and function approximators. Empirically, Value Function Transfer appears to outperform Rule Transfer when both methods are applicable.
Tasks: Ringworld, Knight Joust, 3 vs. 2 Keepaway, and 4 vs. 3 Keepaway **Learners:** Sarsa **Function Approximators:** Tabular, CMAC, and RBF
Method: Representation Transfer: Complexification, Algorithm 11, Section 7.3
Scenario: Complexification is applicable when the parameterization of a function approximator can change over time and the function approximator exhibits locality (such as in a CMAC or RBF).
Tasks: XOR Keepaway **Learners:** Sarsa **Function Approximators:** CMAC
Method: Representation Transfer: Offline Representation Transfer, Algorithms 12–15, Section 7.3
Scenario: Offline Representation transfer is a very flexible method, but requires additional computation between learning the source and target tasks (although it is relatively small in practice).
Tasks: 3 vs. 2 Keepaway and 4 vs. 3 Keepaway **Learners:** Sarsa and NEAT **Function Approximators:** ANN, CMAC, and RBF

7.4 Comparison of Presented Transfer Methods

Chapters 5–7 introduced six different TL methods. In this section, we summarize the method's differences, their relative advantages, and the experiments conducted. Table 7.10 can thus be considered a high-level summary of the last three chapters.

While this monograph has introduced many different transfer algorithms, they all share the same fundamental insight: inter-task mappings can be used to reuse knowledge in a novel task with different state variables and actions. Value Function Transfer can be considered the core TL algorithm of this monograph as it is studied the most thoroughly. Subsequent algorithms are introduced in order to expand the applicability of transfer by allowing transfer between additional base RL methods.

With the exception of TIMBREL, all TL methods in this monograph have been applied to tasks within the Keepaway domain, and many use additional domains which show the methods' broad applicability. (TIMBREL was not able to learn Keepaway because Fitted R-MAX, the base RL algorithm used, is not yet able to scale to such difficult tasks.) By sharing a common domain, we can draw some conclusions about the relative efficacy of each method, but the primary difference between TL methods is the situations in which they apply. In many cases, the base RL algorithms chosen for a particular set of tasks will dictate the type of TL algorithm to use.

These last three chapters have shown that inter-task mappings can:

1. successfully enable transfer between tasks with different state variables and actions,
2. be used as the core of multiple TL methods, which enable transfer over different base RL algorithms and function approximators,
3. be used by TL methods to reduce the target task training time and the total training time,
4. apply to many different tasks in multiple domains.

Up to this point in the monograph, we have assumed that all inter-task mappings are provided to the agent, and are correct. In the following chapter, we will introduce a pair of methods that are able to *learn* inter-task mappings, a necessary step for autonomous transfer. We then provide empirical evidence that the mappings learned by both methods are able to successfully increase the speed of learning in a target task.

Chapter 8
Learning Inter-Task Mappings

Chapter 5 introduced inter-task mappings and Value Function Transfer. The subsequent two chapters introduced a series of TL methods with different characteristics, all of which use the same inter-task mapping formulation. In addition to using full inter-task mappings, partial inter-task mappings were successfully used for transfer (Section 6.2.3). Up to this point in the monograph, all inter-task mappings have been provided as input to TL algorithms. However, some domains may be complex enough that it is difficult to intuit such inter-task mappings. Furthermore, an agent acting autonomously could not be provided such information, but would have to discover it.

This chapter introduces a pair of methods to learn inter-task mappings. The first, *Mapping Learning via Classification* [Taylor et al. (2007b)], requires domain knowledge to correctly partition state variables into different objects. It then uses a classification algorithm on data gathered in both tasks to determine which objects in the source task and target task are most similar, creating an inter-task mapping. The second, *Modeling Approximated State Transitions by Exploiting Regression* [Taylor et al. (2008d)] (MASTER), does not need to be provided such domain knowledge. Instead, it creates an approximate transition model and tests different mappings offline against data gathered in both tasks, finding an appropriate inter-task mapping.

8.1 Learning Inter-Task Mappings via Classification

This section introduces *Mapping Learning via Classification*. To discover appropriate inter-task mappings, the agent first observes its transitions in the source and target tasks. Given groupings of state variables plus gathered experience from source and target tasks, supervised learning methods then autonomously identify similarities between state variables and actions in the two tasks (see Figure 8.1). The primary assumption of this method is that state variables can be arranged into task-independent groupings, and that such background knowledge is provided as input.

M.E. Taylor: Transfer in Reinforcement Learning Domains, SCI 216, pp. 181–204.
springerlink.com

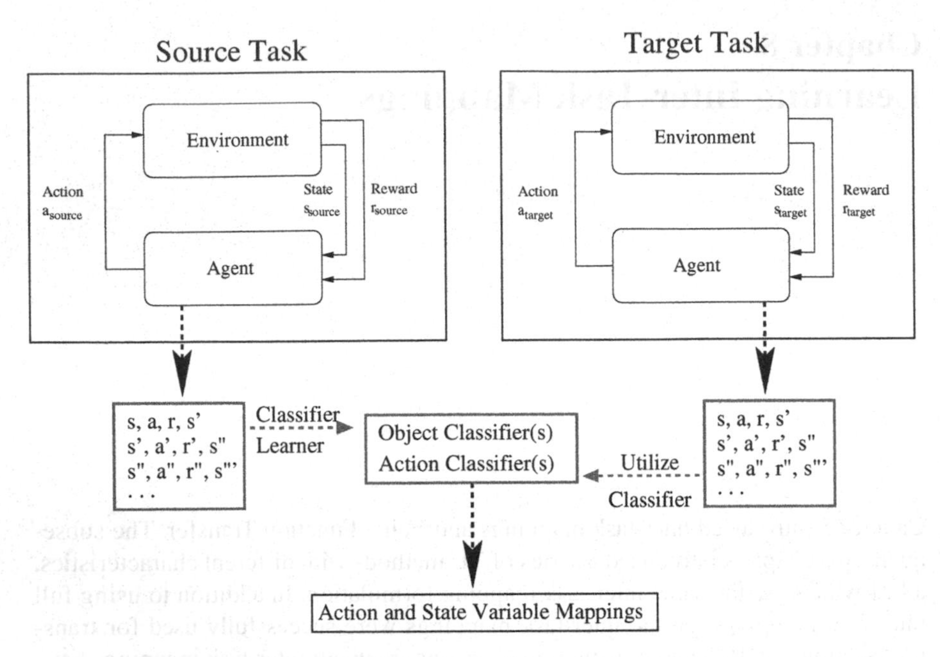

Fig. 8.1 Mapping Learning via Classification has four main steps. First, data is collected from the source task. Second, object and action classifier(s) are trained. Third, data is collected from the target task. Fourth, target task data is classified into source task objects and actions, creating an inter-task mapping.

First, consider the task of finding a mappings between actions in the target task and the source task. The intuition behind this mapping-learning method is that the essence of an action is its effect on the state variables, as determined by the unknown transition function (and possibly the immediate reward, as determined by the unknown reward function). By gathering data in a task, we can learn to classify different actions by how they affect state transitions. The question then becomes how to use such a classifier, given that the state variables and actions can be different in the two tasks.

When learning in the source task, an agent records data samples of the form $(s_{source}, a_{source}, r, s'_{source})$, where r is the immediate reward. s_{source} and s'_{source} are the state of the world (both vectors of k state variables) before and after the agent executes action a_{source}. These samples are used to train classifiers that predict the index of a particular state variable or action, given the rest of the data in the sample. For example, given a state, reward, and next state, classifier C_A predicts the action taken:

$$C_A(s_{source}, r, s'_{source}) = a_{source}.$$

Such a classifier may be used to define a learned inter-task mapping between actions. If the target task had the *same* state variables as the source task, a sample gathered in the target task, $(s_{target}, a_{target}, r, s'_{target})$, could classified by C_A:

$$C_A(s_{target}, r, s'_{target}) = a_{source}$$

and then a_{source} would correspond to a_{target} (i.e., $\chi_A(a_{target}) = a_{source}$).

However, because s_{target} and s'_{target} may have different state variables than s_{source} and s'_{source}, we cannot use target task states as inputs to a classifier trained on source task states. To address this problem, we leverage domain knowledge and do not train classifiers on the full state, but on *subsets* of state variables that define a particular object. Specifically, rather than training an action classifier with $2k+1$ inputs, we will train multiple action classifiers, one per object type.

Suppose that there are T object types that define groups of state variables in a domain. For example, a logistics domain might have two object types: $T = \{truck, location\}$. If there are two trucks and two locations in the source task, then each state variable can be associated with a particular object in $\{truck_A, truck_B, loc_A, loc_B\}$. Instead of learning a single action classifier, C_A, we learn a separate $C_{A,t}$ for each $t \in T$:

$$C_{A,t}(s_{i,source}, r, s'_{i,source}) = a_{source}$$

where $s_{i,source}$ contains the state variables associated with object i of type t in the source task. Thus the inputs for $C_{A,truck}$ will be state variables associated with either $truck_A$ or $truck_B$. Each recorded data tuple (s, a, r, s') thus produces multiple training examples, one for each object described by the state.

Once trained, such classifiers can be used to define χ_A. Each object j of type t in each sample gathered in the target task is input to the relevant $C_{A,t}$:

$$C_{A,t}(s_{j,target}, r, s'_{j,target}) = a_{source}.$$

Each classifier's output is interpreted as a "vote" for a correspondence between a_{target} and a_{source}; $\chi_A(a_{target})$ is set to be the action in the source task with the most votes. Continuing our example, a state in the target task can be partitioned so that the state variables $s_{j,target}$ and $s'_{j,target}$ that describe $truck_A$ in the target task are classified by $C_{A,truck}$, which counts as a vote for an a_{source} similar to the observed a_{target}. Likewise, the state variables corresponding to $truck_B$ are classified by $C_{A,truck}$ to produce a second vote.

We define a similar mapping between state variables, χ_X, by training classifiers to predict which object i is described in the input. Hence, we learn a separate $C_{X,t}$ for each of the t object types:

$$C_{X,t}(s_{i,source}, r, s'_{i,source}) = i.$$

Once trained, these classifiers can be used to define χ_X. Each object j of type t in each sample gathered in the target task is input to the relevant $C_{X,t}$:

$$C_{X,t}(s_{j,target}, r, s'_{j,target}) = i.$$

Again, each classifier's output is interpreted as a "vote" for a correspondence between object j in the target task and object i in the source task, and $\chi_X(s_{j,target})$ is defined by winner-take-all voting. In the logistics example, this could correspond to a vote to classify the target task object $Truck_C$ as the source task object $Truck_A$.

Note that if an appropriate action mapping were already known, each $C_{X,t}$ could utilize χ_A to classify data in the form:

$$C_{X,t}(s_{j,target}, r, s'_{j,target}, \chi_A(a_{target})) = i.$$

Likewise, a given χ_X could be leveraged to learn χ_A. If one of the two classification tasks prove easier to learn, or one of the mappings is given but not the other, one inter-task mapping can be bootstrapped to learn the other.

Notice that this learning method, and the resulting inter-task mapping, are independent of the particular base RL algorithm used and are compatible with all TL algorithms introduced in this monograph. This mapping-learning method instead relies on the assumption of task-independent state variable groupings, the agent's ability to collect experience in both tasks, and an appropriate classification technique to find similarities.

8.1.1 Learning Keepaway Inter-Task Mappings

As discussed in Section 8.1, when learning an inter-task mapping, domain knowledge is used to define how the state space should be semantically partitioned (i.e., how a given state variable should be assigned to some object). In Keepaway, it is natural to divide the state space into keepers, with four state variables each, and takers, defined by two state variables (see Table 8.1), similar to previous transfer research in this domain [Soni and Singh (2006)]. We define the state before an action to be the state perceived by the keeper with the ball (the only agent learning at any given time), and the next state to be the state of the world as perceived from that same keeper after the action has successfully finished (i.e., the next time any keeper can select a macro-action).

To learn the inter-task mapping we train three classifiers using JRip, an implementation of RIPPER [Cohen (1995)] included in Weka [Witten and Frank (2005)]. We selected JRip because it learns quickly and produces human understandable rules, but other classification methods in Weka had comparable results in informal experiments. The three classifiers learned are:

1. $C_{keeper}(s_k, r, s'_k) =$ Source Keeper
2. $C_{taker}(s_t, r, s'_t) =$ Source Taker
3. $C_{action}(s_{3vs2}, r, s'_{3vs2}) =$ Source Action

where s_k and s'_k are the subsets of state variables used to represent a single keeper (an object of type *keeper*) before and after an action has executed; s_t and s'_t describe a taker (an object of type *taker*); s_{3vs2} and s'_{3vs2} describe an entire 3 vs. 2 Keepaway state (three keeper objects and two taker objects); and r is the Keepaway reward accrued between actions (the number of timesteps elapsed).

Table 8.1 The Keepaway state is partitioned into individual keepers and takers

Keepaway State Variable Assignments

Keeper State Variables
$dist(K_n, K_1)$ $dist(K_n, C)$ Min($dist(K_n, T_1)$, $dist(K_2, T_2)$) Min($ang(K_n, K_1, T_1)$, $ang(K_n, K_1, T_2)$)
Taker State Variables
$dist(T_n, K_1)$ $dist(T_n, C)$

Consider using a single $(s_{3vs2}, r, a, s'_{3vs2})$ tuple recorded in 3 vs. 2 used to train these three classifiers. This tuple yields two data points[1] for training C_{keeper}:

- (s_{k_2}, r, s'_{k_2}), label = 2
- (s_{k_3}, r, s'_{k_3}), label = 3

where s_{k_2} and s_{k_3} are the state variables corresponding to keepers 2 and 3. Similarly, this tuple will produce two training examples for C_{taker} (one per taker) and one example for C_{action} which includes the action that was executed.

Once trained, the classifiers are able to process data gathered from agents and label which keeper, taker, or action in the source task it is most similar to. Data recorded from 3 vs. 2 is split into training and test sets to verify correctness of the classifiers on the source task. The three classifiers were able to correctly classify source players and actions, as tested with cross-validation. We then utilize these classifiers to learn a mapping by applying them to data gathered in the target task.

In the target task we again assume that the state variables associated with keepers and takers can be identified. Each $(s_{4vs3}, r, a, s'_{4vs3})$ tuple recorded in the target task produces data for 3 keepers and 3 takers (again, assuming that K_1 is trivially identified). C_{keeper} classifies sets of keeper state variables and C_{taker} classifies the taker state variables, constructing χ_X. As in was done previously in χ_X for Keepaway (Table 5.1 in Section 5.1), two 4 vs. 3 keepers map to a single 3 vs. 2 keeper and two 4 vs. 3 takers map to a single 3 vs. 2 taker.

Using χ_X, the full state in the target task can be reformulated so that only information that was present in the source task is considered by C_{action}. In the target task, each tuple recorded produces four examples for classification because s_{4vs3} may generate four[2] distinct s_{3vs2}'s. χ_A is then constructed by classifying target task data with C_{action}.

[1] We assume that the state variables corresponding to K_1, the keeper with the ball, are known, as the majority of distances are measured relative to it. Learning this keeper would be simple, but complicates the exposition. Likewise, we assume that the hold action is known; it is easy to classify as it is the only action that lasts a single timestep.

[2] s_{3vs2} requires 3 objects of type keeper and 2 objects of type taker. K_1 will always be included in s_{3vs2}, $K_{4,target}$ always maps to $K_{3,source}$, and $T_{1,target}$ always maps to $T_{1,source}$. However, both $K_{2,target}$ and $K_{3,target}$ map to $K_{2,source}$, and both $T_{2,target}$ and $T_{3,target}$ may to $T_{2,source}$ in the learned χ_X (see Table 8.3). Thus the following formulations of s_{3vs2} are possible via χ_X: $(K_1, K_2, K_4, T_1, T_2)$, $(K_1, K_3, K_4, T_1, T_2)$, $(K_1, K_2, K_4, T_1, T_3)$ and $(K_1, K_3, K_4, T_1, T_3)$.

Table 8.2 A representative keeper confusion matrix demonstrates that C_{keeper} has been successfully learned and that it can reasonably classify 4 vs. 3 keepers. This matrix represents an inter-task mapping that would map K_2 and K_3 in the target task to K_2 in the source task, and K_4 in the target task to K_3 in the source task.

Applying a Trained C_{keeper} to 4 vs. 3 Keepaway

	Source K_2	Source K_3
Target K_2	286	17
Target K_3	234	69
Target K_4	37	266

Table 8.3 This table describes three inter-task mappings for Keepaway. Hand-coded inter-task mappings require the most information to construct, partial inter-task mappings require less information (see Section 6.2.3), and the Learned mappings require only information about assignments from state variables to objects. Items in **bold** differ from the hand-coded mapping.

Keepaway Inter-task Mappings: Hand-coded, Partial, and Learned

Target Task Object	Hand-coded $\mathcal{X}_X$	Partial $\mathcal{X}_X$	Learned $\mathcal{X}_X$
K_1	K_1	K_1	K_1
K_2	K_2	K_2	K_2
K_3	K_3	K_3	K_2
K_4	K_3	**none**	K_3
T_1	T_1	T_1	T_1
T_2	T_2	T_2	T_2
T_3	T_2	**none**	T_2
Target Task Action	Hand-coded $\mathcal{X}_A$	Partial $\mathcal{X}_A$	Learned $\mathcal{X}_A$
Hold	Hold	Hold	Hold
Pass K_2	Pass K_2	Pass K_2	Pass K_2
Pass K_3	Pass K_3	Pass K_3	**Pass K_2**
Pass K_4	Pass K_3	**none**	Pass K_3

For our experiments, we collected 1,000 tuples $(s_{3vs2}, r, a_{3vs2}, s'_{3vs2})$ from 3 vs. 2 Keepaway and 100 $(s_{4vs3}, r, a_{4vs3}, s'_{4vs3})$ tuples from 4 vs. 3 Keepaway. Note that collecting the tuples from the source task is "free," assuming that at least that many tuples were experienced during training. Recording the 100 tuples in 4 vs. 3 took about a minute of simulated time, which was negligible compared to the hours later spent training. A representative confusion matrix for the target task keepers is shown in Table 8.2. The learned transfer functional is constructed via a winner-take-all scheme and the resulting mappings are shown in Table 8.3.

8.1.2 Learning Server Job Scheduling Inter-Task Mappings

In SJS, state variables are naturally grouped by job type. A classifier C_{Job} takes as input four state variables that specify job counts for a particular job type (that is, for job type j, the number of jobs aged: 1-50 time steps, 51-100 time steps, 101-150 time steps, and 151-200 time steps). s' is defined to be the state of the world immediately after a job has been removed but before the next job is added to the processing queue. Thus the two classifiers to learn χ_X and χ_A are defined as:

1. $C_{Job}(s_{job}, r, s'_{job}) =$ Source Job Type
2. $C_{Action}(s_{job}, r, jobType_{source}, s'_{job}) =$ Source Action

where *jobType* is the label of the source job type for s_{job}.

As in Keepaway (Section 8.1.1), we first learn χ_X by learning the state variable correspondence between the source and target tasks. C_{Job} is trained with source task agent experience, where every recorded tuple produces two data, one for each job type. Then data from the target task, four for each tuple, is used to map the four target job types into the two source job types. C_{Action} is trained on source experience and two data are generated for each action taken. When using C_{Action} in the target task, since there are four job types, there are four data to classify for each recorded action in the environment. Because actions in SJS are defined as removing a job from a particular job type, we utilize χ_X when using C_{Action}: $C_{Action}(s_{job}, r, \chi_X(jobType_{target}), s'_{job}) = A_{source}$. Thus C_{Action} is used to construct χ_A.

In our experiments, when using 10,000 tuples from the source task and 10,000 tuples from the target task, the learned inter-task mappings for Server Job Scheduling were identical to the hand-coded inter-task mappings (in Section 6.2.1), suggesting that this approach to learning inter-task mappings is effective. On average, 50 episodes (one tenth of a generation) in the target tasks were used to collect data to learn the mappings, which is short when compared to total training times in the target task (on the order of 10,000 episodes).

8.1.3 Learned Inter-Task Mapping Results

In this section we compare NEAT learning times in Keepaway when learning without transfer to using one of the three inter-task mappings in conjunction with Policy Transfer (full, partial, or learned mappings, as listed in Table 8.3).[3] There are two possible conditions for success. One measure of success would be to learn a mapping that was identical, or very similar, to the hand-coded mapping. This would suggest that the hand-coded inter-task mapping was well designed and that Mapping Learning via Classification can successfully learn appropriate mappings. A second measure of success would be to learn a mapping that, when used for transfer, achieved the similar (or perhaps better) performance than when a hand-coded mapping was used for transfer. This measure allows for sub-optimal hand-coded mappings and considers the mapping learning a success if the inter-task mapping produced performs well in practice. Experiments in this section show that the SJS mappings learned are identical to the hand-coded mappings, satisfying both of the above goals. The learned Keepaway mappings are not identical to the hand-coded

[3] Although any of the TL methods introduced in this monograph could be used, we choose to test Policy Transfer with NEAT so that the learned mappings can be compared to the partial-mappings previously introduced (see Section 6.2.3).

mappings, but are qualitatively similar and achieve similar transfer performance, which also satisfies both of the above goals.

As done in Section 6.2, which introduced Policy Transfer, a population of policies is first trained in 3 vs. 2 Keepaway. The entire population is then transferred into 4 vs. 3 via Policy Transfer in conjunction with one of the mappings. To compare learning with and without transfer we set a target threshold performance and measure how much experience the keepers require to reach the threshold (following the experimental methodology in Section 6.2).

Figure 8.2 shows the training time each method required to reach threshold performances of 7.0 to 8.5 seconds. Five learning curves were generated by averaging over 10 independent runs using four learning methods: learning without transfer, using a hand-coded mapping after training for 5 generations of 3 vs. 2 or 10 generations of 3 vs. 2, using partial mappings, and using learned mappings after training for 5 generations of 3 vs. 2. The differences between transfer with learned mappings and no transfer are statistically significant at over half of the points graphed.

When considering the total training time, learning curves in Figure 8.2 which use transfer are shifted up by the amount of time spent training in the source task. Figure 8.3 shows the target and total training times needed to reach a target threshold of 8.5 seconds. The differences between the total training times and no transfer for all four inter-task mappings are statistically significant for roughly half of the target threshold times shown in Figure 8.2.

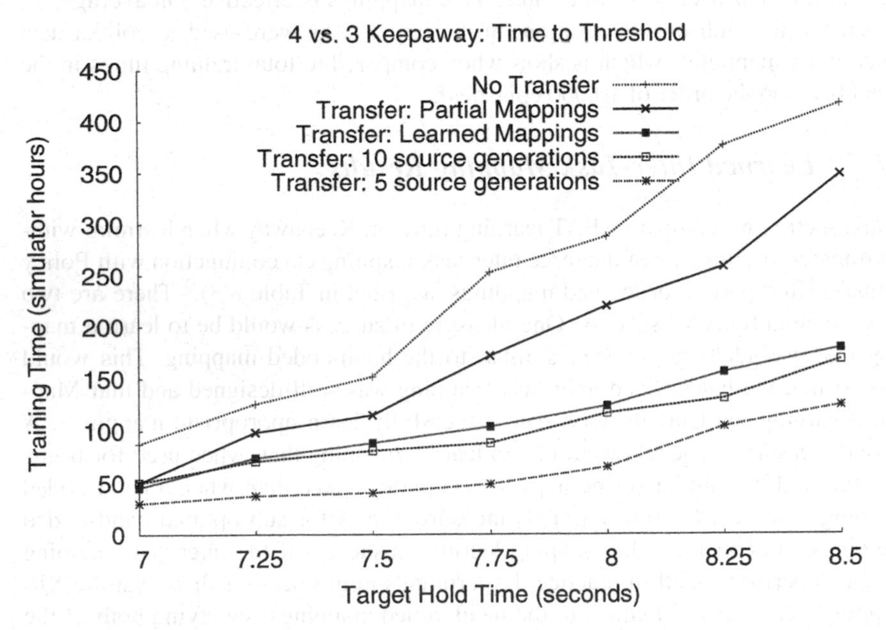

Fig. 8.2 Policy Transfer successfully reduces the average training time needed to reach a given performance level relative to learning without transfer. This graph is similar to Figure 6.12 in Section 6.2.3, but includes transfer results that transfer from 5 source task generations with learned inter-task mappings.

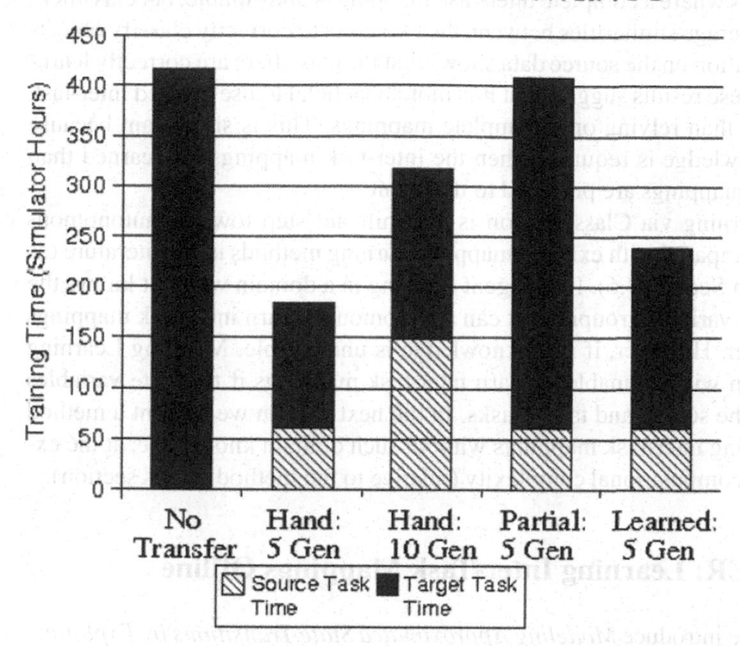

Fig. 8.3 This chart shows the average amount of 4 vs. 3 Keepaway (target) training time and 3 vs. 2 Keepaway (source) training time needed to reach a target performance of 8.5 seconds. Learning without transfer is compared to using Policy Transfer with: hand-coded mappings after 5 and 10 source task generations, partial mappings after 5 source task generations, and learned mappings after 5 source task generations.

In Server Job Scheduling, the learned and hand inter-task mappings are identical. No new results need to be run, as the performance of the learned mappings are the same as the performance of the hand-coded mappings (see Figure 6.8 in Section 6.2.1).

In summary, these results show that Learning Mappings via Classification can successfully learn inter-task mappings. In SJS, the learned and hand-coded mappings were identical. In Keepaway, the learned mappings were very similar to the hand-coded mappings: one keeper, one taker, and one action were mapped differently than the hand-coded mapping, but they were mapped to source task objects and actions that were similar to the hand-coded mapping (see Table 8.3). Furthermore, the learned Keepaway mapping was successfully used by Policy Transfer to improve learning in the target task.

8.1.4 Discussion

In our experiments we found that learned inter-task mappings were able to outperform partial mappings because the learned transfer mappings were more similar to the hand-coded functionals than the partial mappings were to the hand-coded mappings. In domains where a complete inter-task mapping is unavailable, but classification is able to leverage similarities between the two tasks to correctly classify objects (i.e., cross-validation on the source data shows that the classifiers are correctly learning concepts), these results suggest that it is more beneficial to use learned inter-task mappings rather than relying on incomplete mappings. This is significant because less domain knowledge is required when the inter-task mappings are learned than when inter-task mappings are provided to the agent.

Mapping Learning via Classification is a significant step towards autonomous transfer when compared with existing mapping learning methods in the literature (as discussed later in Section 3.6). If the agent is acting in a domain where it knows the appropriate state variable groupings, it can autonomously learn inter-task mappings to enable transfer. However, if such knowledge is unavailable, Mapping Learning via Classification will be unable to learn inter-task mappings if the state variables are different in the source and target tasks. In the next section we present a method capable of learning inter-task mappings without such domain knowledge, at the expense of higher computational complexity (relative to the method in this section).

8.2 MASTER: Learning Inter-Task Mappings Offline

In this section we introduce *Modeling Approximated State Transitions by Exploiting Regression* (MASTER), a method for learning an inter-task mapping without domain knowledge. As in the case of Mapping Learning via Classification, the inter-task mappings learned by MASTER can be used in conjunction with any base RL method or any of the TL methods presented in Chapters 5–7. MASTER can successfully learn both the action mapping, χ_A, and the state variable mapping, χ_X, from data collected in the source and target tasks. This section focuses on a high-level description while implementation-level details will be specified in Section 8.2.1 in the context of specific transfer experiments.

Our domain-independent method for constructing inter-task mappings is summarized in Algorithm 16. We consider five distinct phases (see Figure 8.4):

1. Lines 1–3 represent training in the source task. Any learning method can be used that is capable of utilizing inter-task mappings for transfer (e.g., KBKR [Maclin et al. (2005)], Sarsa, or NEAT). The type of knowledge saved in the data structure D_{source} will depend on which RL algorithm is used for source task learning.
2. Lines 4–5 show the agent(s) exploring in the target task without learning. Experiments suggest that only a relatively small amount of data is needed (see Section 8.2.3).
3. A one-step transition model for the target task, M_{target}, is learned on line 6. As discussed in Section 8.2.1, our experiments utilize neural network function approximation in the Weka machine learning package, but we expect other prediction methods to also perform well. Note that the error calculation $(M_{target}(s,a) - s')$ is a vector operation and is computed per state variable. Such an error

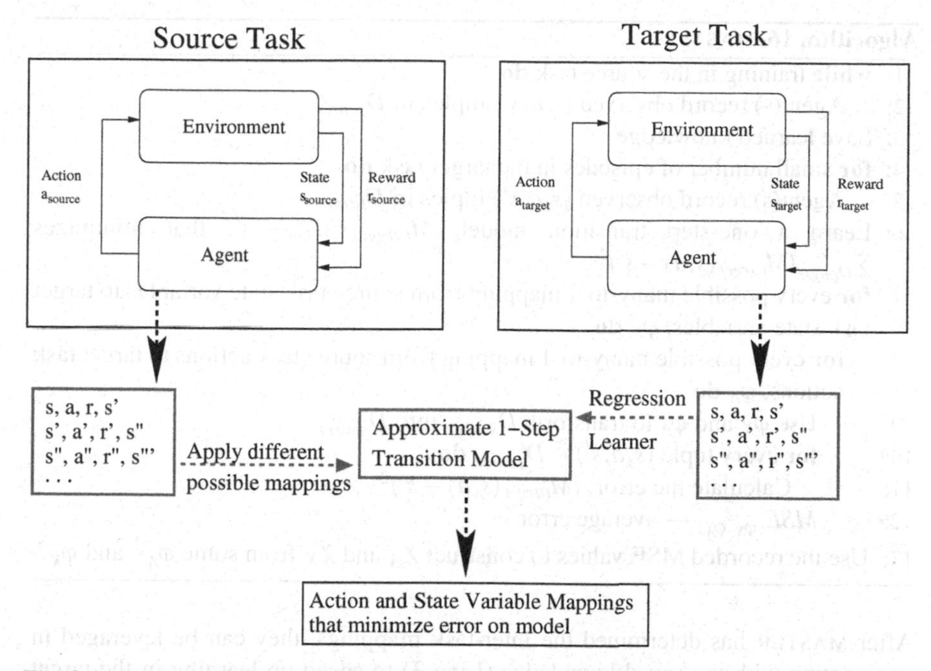

Fig. 8.4 MASTER has five main steps. First, data is collected from the source task. Second, a small amount of data is collected from the target task. Third, an approximate regression model is learned from the target task data. Fourth, different possible mappings are applied to the source task data offline and tested against the model. Fifth, the state variable and action mappings that minimize the error on the model are returned.

definition implicitly assumes that the state variables can be scaled so that they are weighted equally, and that a Euclidean metric is an appropriate measure of state similarity (for both discrete and continuous state variables).

4. Lines 7–12 examine different ways of mapping the source task data into the target task using inter-task mappings φ_X and φ_A.[4] When considering target tasks that have more state variables and/or actions than the source task, this is typically a many-to-one mapping. Each possible mapping is tested and its appropriateness is determined by how well it matches the learned model.
5. Lastly, the agent constructs the inter-task mapping from the tested mappings (line 13). Note that the inter-task mapping maps target task data to source task data, while the agent had been testing different mappings from source task data into the target task. Details of this step will be discussed in Section 8.2.1, but the intuition is that if a there is a single best mapping, it should be used. If there are a number of candidate mappings that have very similar MSEs, they can be combined in a mixture weighted by their inverse errors.

[4] The inter-task mappings may be, by definition, defined either from the target to source or from the source to target. In this monograph χ has been used to denote mappings from the target to source, and thus we use φ when referring to state variable and mappings from the source task to the target task.

Algorithm 16. MASTER

1: **while** training in the source task **do**
2: Agent(s) record observed (s,a,s') tuples in D_{source}
3: Save learned knowledge
4: **for** small number of episodes in the target task **do**
5: Agent(s) record observed (s,a,s') tuples in D_{target}
6: Learn a one-step transition model, $M_{target}(s,a) \mapsto s'$, that minimizes $\sum_{D_{target}}(M_{target}(s,a) - s')^2$
7: **for** every possible many-to-1 mapping from source task state variables to target task state variables, φ_X **do**
8: **for** every possible many-to-1 mapping from source task actions to target task actions, φ_A **do**
9: Use φ_X and φ_A to transform D_{source} into D'_{source}
10: **for** every tuple $(s,a,s') \in D'_{source}$ **do**
11: Calculate the error: $(M_{target}(s,a) - s')^2$
12: $MSE_{(\varphi_X,\varphi_A)} \leftarrow$ average error
13: Use the recorded MSE values to construct χ_A and χ_X from some φ_A^{-1} and φ_X^{-1}

After MASTER has determined the inter-task mappings, they can be leveraged in conjunction with the saved knowledge (Line 3) to speed up learning in the target task using one of the TL methods for RL tasks.

The key insight of this method is that MASTER is able to propose all possible methods and then score them by analyzing them offline (i.e., without requiring more samples from the environment), similar to the approach taken by [Mihalkova et al. (2007)] to map predicates between different relational domains. Such analysis, lines 7–12, is exponential in the number of state variables and actions. Scaling MASTER to tasks with a large number of state variables or actions would require some type of heuristic. For instance, rather than an exhaustive search, a hill-climbing method could be used to find a good mapping (e.g., a variant on Powell's Method [Powell (1964)]). It is worth emphasizing that this search affects only computational complexity. In this work we attempt to learn an inter-task mapping so that the sample complexity of the target task is reduced – reducing the computational complexity is not our primary concern, as CPU cycles are generally cheap when compared to collecting data from a fielded agent.

There are a number of model-learning methods for RL tasks (e.g., KBRL [Ormoneit and Sen (2002)]), but such methods do not generally scale to large tasks with continuous state variables, which are of particular interest to agents acting in real-world tasks. Such methods generally attempt to model a task in order to perform dynamic programming offline. In MASTER, we instead only need to learn an *approximate* model that allows us to find similarities between state variables and actions in two tasks. Since we typically expect relatively large differences in the transition model of an MDP when state variables and actions are changed, the error due to poor modeling is less critical then when a model is used for dynamic programming. This relaxed requirement allows us to use a simple regression method, which may be used on tasks with continuous state variables and requires relatively little data for model learning.[5]

8.2.1 *MASTER: Empirical Validation*

In this section we demonstrate how MASTER can learn inter-task mappings from the Standard 3D Mountain Car to the Standard 2D Mountain Car. We then discuss a number of experiments that illustrate how MASTER is able to achieve a significant speed-up in the target task with limited source task data and compare the results of our algorithm with Mapping Learning via Classification. Lastly, we demonstrate how MASTER can evaluate mappings between multiple source tasks and help to select an appropriate source task for transfer.

8.2.1.1 Applying MASTER to Mountain Car

As discussed in Section 4.1.3, Mountain Car tasks can be learned with Sarsa and CMAC function approximation. In this section we first use MASTER to learn the inter-task mappings between two mountain car tasks. We then use the learned mappings to transfer between the Standard 2D Mountain Car and Standard 3D Mountain Car tasks.

In order to use MASTER to learn an inter-task mapping between Mountain Car tasks, the agent first trained in Standard 2D Mountain Car for 100 episodes using Sarsa(λ) while saving the observed (s_s, a_s, s_s') transitions (Algorithm 16, lines 1–3). The agent then executed actions randomly in Standard 3D Mountain Car for 50 episodes, recording the observed (s_t, a_t, s_t') transitions.

To learn the one-step transition model (Algorithm 16, line 6), we used the Weka package (version 3.4.6) to train neural networks (i.e., multi-layer perceptrons). While we experimented primarily with neural networks for building the 1-step model, we expect that other function approximators could work equally well. After trying four different parameter settings in informal experiments, we used Weka's default settings, except the number of hidden nodes is set to eight and the number of training epochs is set to 5000. For each Standard 3D Mountain Car trail, we trained a separate neural network for each (action, state variable) pair to learn an approximate model of the transition function, resulting in a total of 20 trained neural networks. Each network modeling the target task data for Standard 3D Mountain car had 4 inputs, one for each state variable, 8 hidden nodes, and a single output that would predict a single state variable's next value.

Once the models are learned, the agent iterates over all possible state variable and action mappings. For instance, it would sequentially try mapping x in the source task to each of $\{x,\ y,\ \dot{x},\ \dot{y}\}$ in the target task. Likewise, the source action Left would be mapped to the target actions {Neutral, West, East, South, North}. The agent then transforms the recorded 2D Mountain Car data using each of these 240

[5] If a significant amount of data from the target task were needed to learn a transition model, relative to the amount of data needed to learn the target task, the time spent gathering data to learn an inter-task mapping could easily outweigh any savings gained by transfer.

mappings (16 state variable mappings $\times$ 15 action mappings). For instance, consider a recorded source task tuple $(x, \dot{x}, \texttt{Left})$, the state variable mapping $\varphi_X : x_s \mapsto \{x_t, y_t\}, \dot{x}_s \mapsto \{\dot{x}_t, \dot{y}_t\}$, and the action mapping $\varphi_A : \texttt{Left}_s \mapsto \texttt{West}_t, \texttt{Right}_s \mapsto \{\texttt{Neutral}_t, \texttt{East}_t, \texttt{South}_t, \texttt{North}_t\}$. Using these mappings, the tuple will be transformed into $(x, x, \dot{x}, \dot{x}, \texttt{West})$. Each transformed tuple is used as input to the neural networks for the relevant target task action. In the above example, the set of four neural networks trained on the target task action West would predict the next state that the agent observes. The output from each neural network is compared with the next state the agent observed in the recorded source task data, and the error over all the transformed source task data is used to calculate the MSE for the mapping.

Table 8.4 summarizes results of a representative trial when evaluating the 240 mappings. For this domain, one state variable mapping is significantly better than all others, both when averaged across all action mappings, or when the best action mapping is considered for each possible state variable mapping. This state variable mapping is fairly intuitive: the position state variable in the Standard 2D Mountain Car maps to both position variables in 3D Mountain Car, and the velocity state

Table 8.4 This table shows the resulting MSE when using different state variable mappings. Each row shows a different mapping where the source task variables in the row are mapped to the target task variables at the head of the column (i.e., x, x, $\dot{x}$, $\dot{x}$ maps variable x_s to x_t, x_s to y_t, $\dot{x}_s$ to $\dot{x}_t$, and $\dot{x}_s$ to $\dot{y}_t$). The Avg. MSE column shows the MSE averaged over all possible action mappings for each row's state variable mapping. The Best MSE column shows the MSE for each row's state variable mapping when using the action mapping with the lowest MSE. Both metrics show that the state variable mapping in bold is significantly better than all other possible state variable mappings.

State Variable Mappings Evaluated

$x\ y\ \dot{x}\ \dot{y}$	Avg. MSE	Best MSE
$x\ x\ x\ x$	0.0384	0.0348
$x\ x\ x\ \dot{x}$	0.0246	0.0228
$x\ x\ \dot{x}\ x$	0.0246	0.0227
$x\ x\ \dot{x}\ \dot{x}$	**0.0107**	**0.0090**
$x\ \dot{x}\ x\ x$	0.0451	0.0406
$x\ \dot{x}\ x\ \dot{x}$	0.0385	0.0350
$x\ \dot{x}\ \dot{x}\ x$	0.0312	0.0289
$x\ \dot{x}\ \dot{x}\ \dot{x}$	0.0245	0.0225
$\dot{x}\ x\ x\ x$	0.0451	0.0406
$\dot{x}\ x\ x\ \dot{x}$	0.0312	0.0290
$\dot{x}\ x\ \dot{x}\ x$	0.0384	0.0350
$\dot{x}\ x\ \dot{x}\ \dot{x}$	0.0245	0.0226
$\dot{x}\ \dot{x}\ x\ x$	0.0516	0.0463
$\dot{x}\ \dot{x}\ x\ \dot{x}$	0.0450	0.0407
$\dot{x}\ \dot{x}\ \dot{x}\ x$	0.0450	0.0407
$\dot{x}\ \dot{x}\ \dot{x}\ \dot{x}$	0.0383	0.0350

Table 8.5 This table shows the MSE found when a source task action is mapped into a target task action. All experiments in this table use the same state variable mapping.

Action Mappings Evaluated

Target Task Action	Source Task Action	MSE
Neutral	Left	0.0118
Neutral	Neutral	0.0079
Neutral	Right	0.0103
West	Left	0.0095
West	Neutral	0.0088
West	Right	0.0127
East	Left	0.0144
East	Neutral	0.0095
East	Right	0.0089
South	Left	0.0099
South	Neutral	0.0093
South	Right	0.0135
North	Left	0.0136
North	Neutral	0.0100
North	Right	0.0100

variable in the Standard 2D Mountain Car maps to both velocity state variables in the 3D Mountain Car.

In Table 8.5 we focus on the best state variable mapping and show the MSE for each of the different possible action mappings. With the exception of Neutral, each task action has two source task actions with very similar error. This effect is caused by the doubling of state variables and actions when using Standard 2D Mountain Car data as input to a Standard 3D Mountain Car model. When using the state variable mapping described above, $\dot{x}_s$ is mapped to both $\dot{x}_t$ and $\dot{y}_t$. Consider saved source task data for the action Right. Right in the source task will cause $\dot{x}_s$ to increase. East in the target task will likewise cause $\dot{x}_t$ to increase, but will not affect $\dot{y}_t$. Because $\dot{x}_s$ has been mapped to both of these state variables, one value will be modified as the target task model expects for the action East, but the other value will not. Intuitively, an appropriate action mapping would map both Right and Neutral from the source task to the action East in the target task. Because there is no clear single best 1-1 mapping, we weight the different action mappings by the inverse of their measured MSE. Such a method will allow us to map multiple actions from the source task into the target task, weighted by their relative errors on our model.

Once the agent learns the mappings φ_X and φ_A, MASTER constructs the inter-task mappings χ_X and χ_A by taking the inverse of these mappings. We then use a transfer method which is very similar to that of Q-Value Reuse (Section 6.1). In this transfer method, the agent saves the 2D CMAC after training on the source task. In the target task, the agent modifies the weights in a 4D CMAC when learning. However, when computing the action-value for a s,a pair, the agent also uses

Algorithm 17. Q-Value Reuse in 3D Mountain Car

1: $x, y, \dot{x}, \dot{y} \leftarrow$ agent's current state
2: $a_t \leftarrow$ action to evaluate
3: **for** each source task action a_s **do**
4: $\quad SUM \mathrel{+}= MSE_{a_t,a_s}$
5: **for** each source task action a_s **do**
6: $\quad Q(s,a_t) \mathrel{+}= Q_{2dCMAC}(x,\dot{x},a_s) \times \mathrm{SUM}/(\mathrm{MSE}_{a_t,a_s})$
7: $\quad Q(s,a_t) \mathrel{+}= Q_{2dCMAC}(y,\dot{y},a_s) \times \mathrm{SUM}/(\mathrm{MSE}_{a_t,a_s})$
8: $Q(s,a_t) \mathrel{+}= Q_{4dCMAC}(x,y,\dot{x},\dot{y},a_t)$

the saved 2D CMAC to evaluate the current position. Conceptually, $Q(s_t,a_t) = Q_{4DCMAC}(s_t,a_t) + Q_{2DCMAC}(\chi_X(s_t),\chi_A(s_a))$. However, as mentioned above, our action mapping is not one-to-one. Thus we iterate over all source task actions, multiply each by the inverse of the action mapping's recorded MSE, and then renormalize (see Algorithm 17). The action mappings with the lowest error have the most influence on the value contributed by the source task CMAC. While learning, the target task CMAC's weights are modified by Sarsa(λ) and will allow for an accurate approximation of the action-value function, even though the transferred source CMAC (which remains unchanged) will not be optimal in the target task.

8.2.2 *Transfer from 2D to 3D Mountain Car*

Figure 8.5 shows learning curves in Standard 3D Mountain car, each averaged over 25 independent trails. During each trial the policy is evaluated after each episode without exploration, but such evaluation episodes are used only for evaluation purposes and are not charged to the learning algorithm. To graph the learning curve we average all 25 learning curves for the previous 10 episodes and plot the mean. First, consider the lines "Without transfer" and "Average Both." The Average Both experiments transfer by averaging over all action mappings and all state variable mappings. Such a method can be considered a type of blind transfer – no time or samples are spent learning an inter-task mapping, but the resulting learning curve is much worse than learning without transfer. Evidently, transferring without any consideration of the state and action variable mappings may be quite harmful to learning. However, as is shown by the other transfer experiments, using MASTER to learn these mappings can enable transfer that is quite beneficial.

The line "Transfer: 1/MSE" is generated by transferring from 100 episodes of Standard 2D Mountain Car where the action mapping is weighted by the inverse of its observed MSE in the target task model. Using paired t-tests we find that the 1/MSE transfer curve is statistically significantly better, at the 95% level, than learning without transfer for episodes 2–473.[6] Also included in the graph are three other transfer mappings for comparison. "Hand Coded" uses hand-coded state variable

[6] On the first episode, the agent with transferred knowledge has an average reward of -4640 while the agent learning without transfer has an average reward of -5000, which is not different at the 95% confidence level.

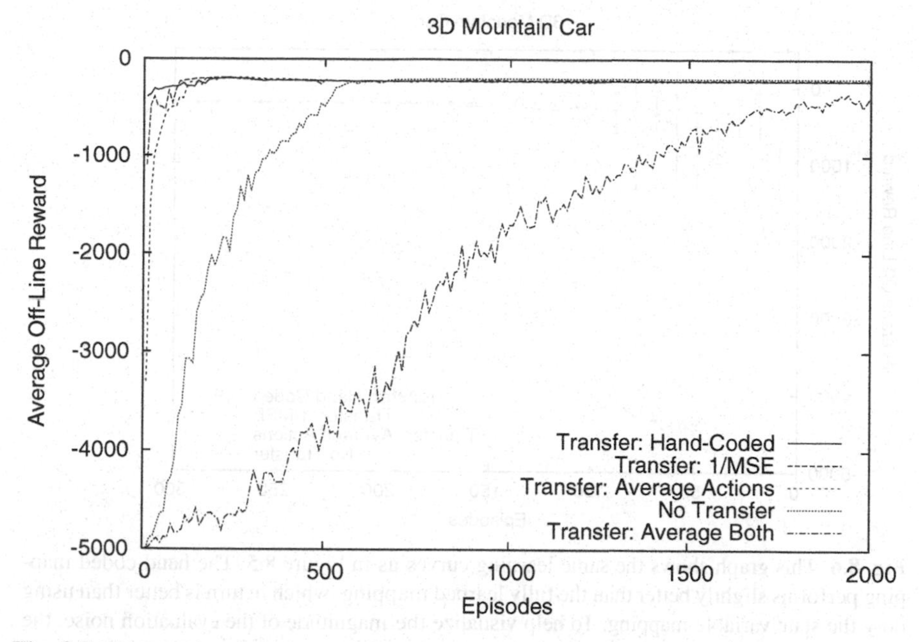

Fig. 8.5 This graph compares transfer with a hand-coded inter-task mappings to: transfer with learned state variable and action mappings, transfer using only a learned state variable mapping, learning without transfer, and transfer with mappings that average over all possible mappings. Figure 8.6 zooms in on the beginning of the same curves. Each learning curve averages 25 independent trials with a 10 episode sliding window.

and action mappings based on our knowledge of the domains. We believe that this learning curve represents the upper bound on transfer for 100 episodes of Standard 2D Mountain Car. It is encouraging that the 1/MSE learning curve quickly converges to the same asymptotic value as the hand coded transfer learner. "Average Actions" performs transfer with the learned state variable mapping but simply averages over all actions. This is equivalent to assuming that all the possible action mappings have the same error and it indicates the importance of using an action mapping. Figure 8.6 shows a magnified version of the graph to better see differences between the different learning curves.

We also tested a final method for weighting the different action mappings. Rather than using all action mappings and weighting by the inverse of the MSE, we selected only the one or two best actions. The learning curve resulting from this method was qualitatively similar to the 1/MSE learning curve and is not shown.

8.2.3 *Reducing the Total Sample Complexity*

The results in Figure 8.5 show that learned source task knowledge can be effectively used with a learned mapping. Thus, if an agent has already trained on Standard 2D

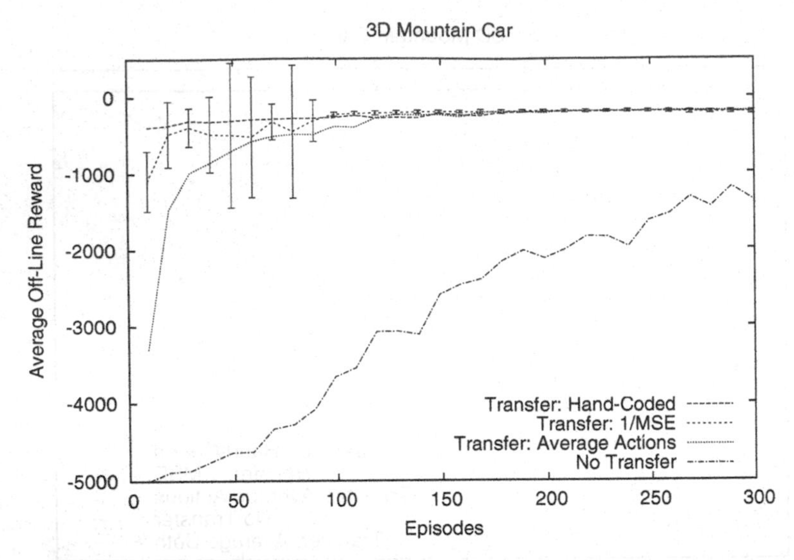

Fig. 8.6 This graph shows the same learning curves as in Figure 8.5. The hand-coded mapping performs slightly better than the fully learned mapping, which in turn is better than using only the state variable mapping. To help visualize the magnitude of the evaluation noise, the 1/MSE transfer curve shows error bars at one standard deviation.

Mountain Car and wants to learn Standard 3D Mountain Car, it likely makes sense to use its past knowledge rather than to learn without it. However, consider a situation where the agent has not trained on Standard 2D Mountain Car and is faced with the Standard 3D Mountain Car task. Should it first train on the 2D task, learn a mapping, and then transfer? Or should it directly tackle the more difficult 3D task?

To help answer the above question, we varied the amount of data used in the source and target task to learn a mapping, as well as how many episodes in the source task used to learn the 2D CMAC's weights. Earlier experiments showed transfer after learning for 100 episodes in the source task, spending 50 episodes collecting data in the target task, and then using MASTER with transfer to learn the target task. We tried using 100, 50, 25, and 10 episodes of source task training, as well as 50, 25, and 10 episodes of target task training. We found that only when we reduce the number of source task episodes to 10 does learning performance in the target task degrade.

Figure 8.7 compares learning Standard 3D Mountain Car without transfer to using transfer. The agent trains for 25 episodes in the source task, collects data for 10 episodes in the target task, uses MASTER to learn the inter-task mappings, weights the action mappings by 1/MSE, and then learns in the target task. Note that the transfer learning curve has been shifted by 35 episodes (the first graphed point is at episode 45, instead of at episode 10) to account for the episodes spent before learning in the target task. A series of paired t-tests show that the difference between learning without transfer and learning with transfer while accounting for all

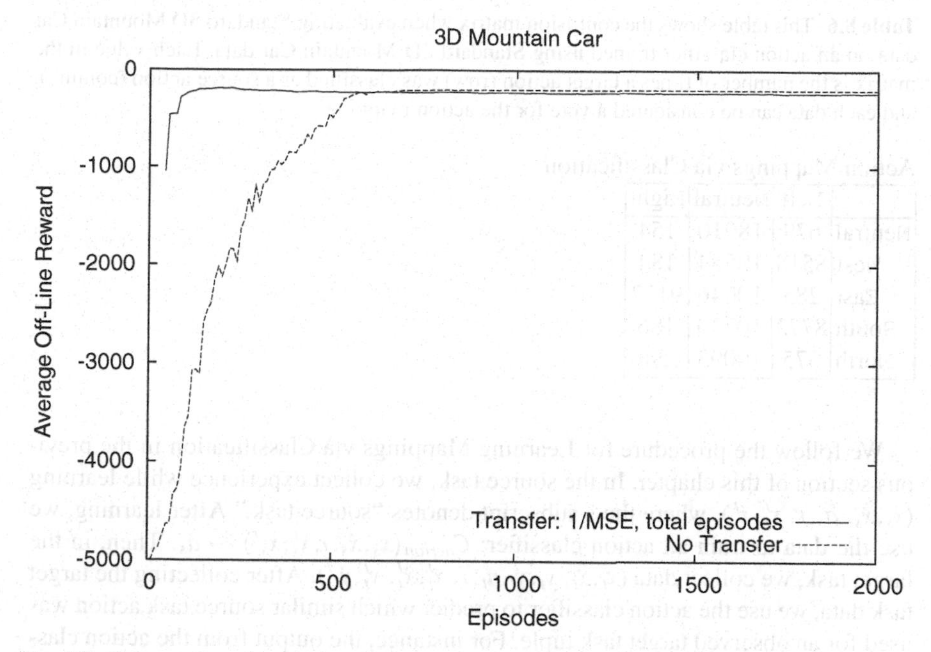

Fig. 8.7 This graph compares learning without transfer to transfer using learned mappings. The transfer learning curve does not start at 0 episodes as it now reflects the total number of episodes used to learn the mappings in the source and target task. This result shows that the total number of episodes to learn a source task, record target task data to learn an inter-task mapping, and then learn in a target task with transfer may less than learning a target task directly. Each learning curve averages 25 independent trials.

episodes used is statistically significantly different at the 95% level from learning without transfer from episodes 36–474. We therefore conclude that for some tasks, it may be in an agent's interest to train first on a simple source task, learn a mapping, and then learn on a target task, rather than learn in the target task directly.[7]

8.2.4 *Comparison to Previous Work*

We would like to compare our method with the previous mapping-learning method, but the 2D and 3D Mountain Car tasks do not easily subdivide into groups of state variables, as was done in Section 8.1. In Mountain Car there is no clear division of "object types." To enable a comparison, we group state variables into (position, velocity) tuples. Our source task will thus have one object, $(x,\dot{x})$, and the target task will have two objects, $(x,\dot{x})$ and $(y,\dot{y})$. A significant amount of information about the relationship between the two tasks has already been encoded in this formulation.

[7] The learning parameters for the Standard 3D Mountain Car task were tuned for learning without transfer. In a different series of experiments, not shown, the transfer learning curves were improved by re-tuning the learning parameters.

Table 8.6 This table shows the confusion matrix when evaluating Standard 3D Mountain Car data on an action classifier trained using Standard 2D Mountain Car data. Each value in the matrix is the number of times a target action (row) was classified as a source action (column), and each data can be considered a vote for the action mapping.

Action Mappings via Classification

	Left	Neutral	Right
Neutral	679	18910	154
West	8518	10554	184
East	285	10046	9177
South	8773	10730	186
North	375	10093	9540

We follow the procedure for Learning Mappings via Classification in the previous section of this chapter. In the source task, we collect experience while learning $(x_s, \dot{x}_s, a_s, r, x_s', \dot{x}_s')$, where the s subscript denotes "source task." After learning, we use the data to train an action classifier: $C_{action}(x_s, \dot{x}_s, r, x_s', \dot{x}_s{}') \mapsto a_s$. Then, in the target task, we collect data $(x_t, \dot{x}_t, y_t, \dot{y}_t, a_t, r, x_t', \dot{x}_t', y_t', \dot{y}_t{}')$. After collecting the target task data, we use the action classifier to predict which similar source task action was used for an observed target task tuple. For instance, the output from the action classifier $C_{action}(x_t, \dot{x}_t, r, x_t', \dot{x}_t{}')$ would give some source task action. The returned source task action is counted as a vote that the target task action associated with this tuple, a_t, is the same as the action returned by the classifier, a_s. Note that no state variable classifier is needed, as there is only one object in the source task. Thus, $(x_t, \dot{x}_t)$ and $(y_t, \dot{y}_t)$ both get mapped to $(x_s, \dot{x}_s)$ because of the knowledge we implicitly gave the agent the state variable grouping. χ_X has been provided by human knowledge, but the classifier is responsible for learning χ_A.

We collected 50 episodes of data in Standard 2D Mountain Car and trained a neural network action classifier with 5 inputs (four state variables and the current reward) to predict the source task action that was taken. The neural network was unable to learn to correctly classify the data until we changed the agent's policy so that the car took each action for 5 successive time steps. By grouping successive states together (i.e., instead of using the state at times t and t+1, we used the state at times t and t+5), the effects of actions outweighed the effects of gravity and we were able to learn to accurately classify source task actions. The action mapping learned is similar to the results of MASTER, as expected (see Table 8.6). If we use this action mapping to learn in Standard 3D Mountain Car (not shown), weighting the different actions by the number of "votes" each mapping received, we find that the target task learning is very similar to our 1/MSE method using both the learned state variable and action mappings described above. The main significance of this result is that it confirms that MASTER is able to find an action mapping similar to that found by an existing learning method, even though significantly less knowledge is required.

8.2.5 *Transfer in Hand Brake Mountain Car*

In this section we examine transfer into the 3D Hand Brake Mountain Car task where an added action sets the velocity to zero on execution, as defined in Section 4.1. First, consider an agent that has previously trained for 500 episodes of 2D Hand Brake Mountain Car. Figure 8.8 compares learning without transfer, learning after transferring only the state variable mapping, and learning after transferring both the state variable and action mapping. This result confirms that MASTER can learn a useful inter-task mapping in this variant of Mountain Car.

Consider an agent that has previously trained on Standard 2D Mountain Car, both with and without a hand brake action. If the agent is now tasked with 3D Hand Brake Mountain Car, it should be able to learn mappings for both tasks and use the learned mappings to intelligently transfer from the source tasks. One option would be to select the source task with learned mappings that had the lowest MSE, which in this case would be Standard 2D Mountain Car with a hand brake action (see Table 8.7). A second option would be to weight the mappings from both tasks by the inverse of their recorded MSEs. Figure 8.9 shows both of these methods outperform transferring only from the 2D Mountain Car without hand brake, as well as outperforming learning the target task without transfer. Interestingly, transferring

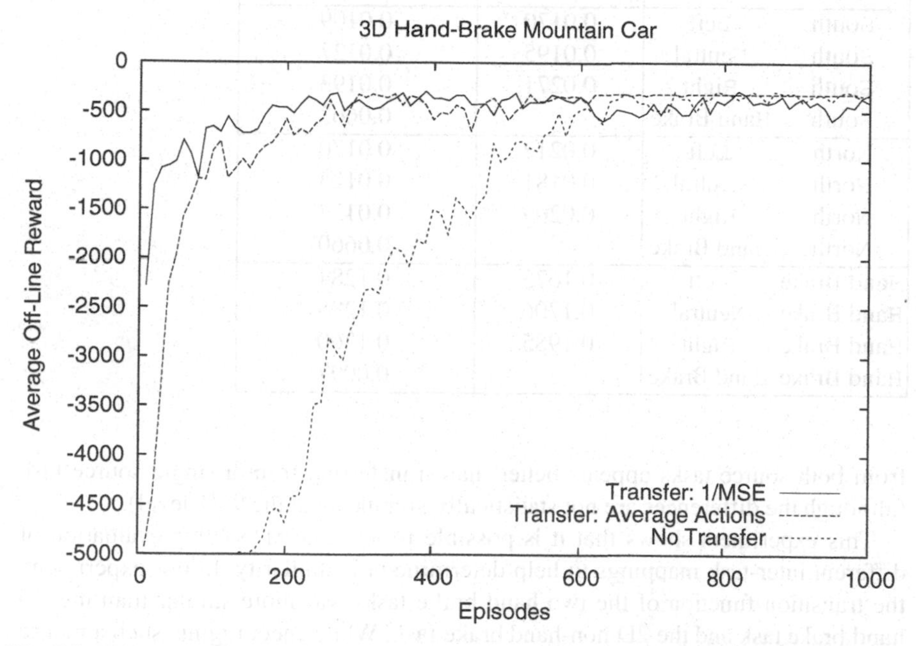

Fig. 8.8 This graph shows that transfer using both learned mappings outperforms both learning without transfer and using only the learned state variable mappings. Each learning curve averages 25 independent trials.

Table 8.7 This table shows a representative mean squared error result when source task actions from Standard 2D Mountain Car (with and without a hand brake action) are mapped into a 3D Hand Brake Mountain Car action. Note that the errors for the 2D hand brake task are lower than the standard 2D task and that no source task action from the standard 2D task maps well to the 3D hand brake action.

Action Mappings for 2 Source Tasks

Target Task Action	Source Task Action	MSE for 2D as Source Task	MSE for 2D Hand Brake as Source Task
Neutral	Left	0.0196	0.0140
Neutral	Neutral	0.0188	0.0113
Neutral	Right	0.0244	0.0162
Neutral	Hand Brake		0.0665
West	Left	0.0180	0.0111
West	Neutral	0.0226	0.0143
West	Right	0.0320	0.0219
West	Hand Brake		0.0678
East	Left	0.0203	0.0162
East	Neutral	0.0175	0.0118
East	Right	0.0217	0.0136
East	Hand Brake		0.0651
South	Left	0.0170	0.0109
South	Neutral	0.0195	0.0127
South	Right	0.0271	0.0193
South	Hand Brake		0.0663
North	Left	0.0212	0.0170
North	Neutral	0.0181	0.0125
North	Right	0.0209	0.0137
North	Hand Brake		0.0660
Hand Brake	Left	0.1673	0.1284
Hand Brake	Neutral	0.1706	0.1285
Hand Brake	Right	0.1985	0.1360
Hand Brake	Hand Brake		0.0097

from both source tasks appears better than transferring from a single source task (although the differences are not statistically significant at the 95% level).

This experiment shows that it is possible to leverage MASTER's evaluation of different inter-task mappings to help determine task similarity. In our experiment, the transition function of the two hand brake tasks was more similar than the 3D hand brake task and the 2D non-hand brake task. While encouraging, such a metric only accounts for the similarity of two tasks' transition functions. If, for instance, the target task's goal state were moved from $(0.5,0.5)$ to $(-1.2,-1.2)$, it is unlikely that transferring from either Standard 2D Mountain Car task would improve learning. In fact, when transferring from such mismatched tasks, it is possible that transfer would

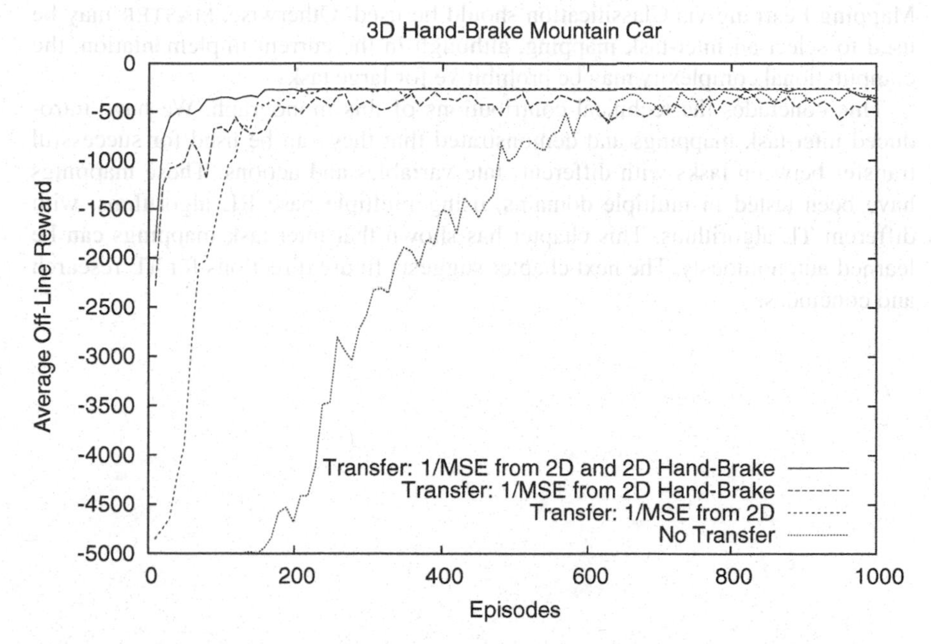

Fig. 8.9 This graph compares learning with transfer from both the 2D and 2D hand-brake tasks (weighted by the inverse of their respective mapping errors), learning with transfer from the 2D hand-brake task alone, learning with transfer from the Standard 2D task, and learning without transfer. Each learning curve averages 25 independent trials.

hurt the learner's performance, relative to learning without transfer. Insulating an agent from the effects of such *negative transfer* is a difficult problem that we leave to future work, along with refining this proposed task similarity metric to account for differences in source and target tasks' reward functions.

8.3 Chapter Summary

This chapter has introduced MASTER, a method for automatically learning a mapping between tasks. We have empirically demonstrated the efficacy of this algorithm on a series of tasks in the Mountain Car domain. These results show that a learned task mappings can effectively increase the speed of learning in a novel target task so that the sample complexity is reduced using transfer, relative to learning without transfer. Additionally, we show an initial approach for leveraging learned inter-task mappings to assist with the problem of appropriate source task selection.

The chapter also introduced Mapping Learning via Classification, which requires relatively little computation to discover a mapping, but requires domain knowledge. In situations where a full mapping is known, the full mapping should be used. If a full mapping is not known but state variables can be partitioned into different object types, or state variables do not change between the source task and the target task,

Mapping Learning via Classification should be used. Otherwise, MASTER may be used to select an inter-task mapping, although in the current implementation, the computational complexity may be prohibitive for large tasks.

This concludes the technical contributions of this monograph. We have introduced inter-task mappings and demonstrated that they can be used for successful transfer between tasks with different state variables and actions. These mappings have been tested in multiple domains, using multiple base RL algorithms, with different TL algorithms. This chapter has shown that inter-task mappings can be learned autonomously. The next chapter suggests future directions for TL research and concludes.

Chapter 9
Conclusion and Future Work

This chapter summarizes the contributions of this monograph and suggests possible enhancements to TL algorithms as well as general open question in transfer learning. Section 9.1 presents a summary of the algorithms introduced by this monograph along with guidelines that suggest when methods are applicable and when one method may be preferable to another. Section 9.2 discusses a series of possible enhancements to the transfer and mapping-learning algorithms presented in this monograph. Section 9.3 suggests situations where the TL methods from this monograph are likely to be effective, as well as discussing the general problem of negative transfer. General open questions in the field of transfer for reinforcement learning are presented in Section 9.4 and Section 9.5 concludes.

9.1 Summary of Monograph Methods

The primary technical contribution of this monograph is to introduce inter-task mappings (Section 5.1), a novel construct that enables transfer between MDPs with different state variables and actions. A second contribution is to introduce six transfer methods (summarized in Table 9.1) that can utilize inter-task mappings to transfer between tasks with different state variables and actions. A third contribution is the introduction of a pair of mapping-learning methods (also in Table 9.1) capable of learning inter-task mappings from observed data via existing machine learning algorithms.

Recall that we motivated the research in this monograph by asking the following question in Chapter 1:

> Given a pair of related RL tasks that have different state spaces, different available actions, and/or different representative state variables,
>
> 1. how and to what extent can agents transfer knowledge from the source task to learn faster or otherwise better in the target task, and
> 2. what, if any, domain knowledge must be provided to the agent to enable successful transfer?

M.E. Taylor: Transfer in Reinforcement Learning Domains, SCI 216, pp. 205–218.
springerlink.com

The take away message from this monograph is that inter-task mappings can successfully enable transfer and thereby significantly improve learning. We have demonstrated that our methods are broadly applicable in different domains, using

Table 9.1 This table summarizes all methods introduced in this monograph and suggests when each method is most useful. Recall that Table 7.10 in Chapter 7 presented a summary of the TL methods, but focuses on experimental results presented, rather than the relative strengths of the different TL methods.

TL Methods Introduced in this Monograph
Method Name: Value Function Transfer, Section 5.2
Scenario: Value Function Transfer is only applicable when both the source task agent and target task agent use TD learning, and both use the same type of function approximation. **When to use:** Results suggest that Value Function Transfer improves learning speed in the target task more effectively than other TL methods and thus should be preferred over other methods, when applicable.
Method Name: Q-Value Reuse, Section 6.1
Scenario: This method again applies when the source and target task agents use TD learning, but agents are not required to use the same function approximator. **When to use:** Transfer with Q-Value Reuse is not as not as effective as Value Function Transfer and may require extra memory due to multiple function approximators, but it becomes more useful if the source task agent and target task agent have different function approximators.
Method Name: Policy Transfer, Section 6.2
Scenario: This method is applicable when the source and target task agents use direct policy search with ANN action selectors. **When to use:** If the target task agents use policy search learning, Policy Transfer is the only applicable TL method introduced by this monograph.
Method Name: TIMBREL, Section 7.1
Scenario: This TL method is applicable when the target task agent uses instance-based RL. The source task agent may use any RL method. **When to use:** TIMBREL is the only method introduced in this monograph that can transfer into a model-learning target task agent.
Method Name: Rule Transfer, Section 7.2
Scenario: Rule Transfer is applicable when the target task agent uses a TD method. **When to use:** If the source task uses an RL method other than value-function learning, but the target task uses value-function learning, Rule Transfer and Representation Transfer are the only applicable methods presented in this monograph. If both the source task agent and the target task agent use value-function learning methods, experiments suggest that Value Function Transfer can provide better learning performance in the target task than Rule Transfer provides.

Table 9.1 *(continued)*

Method Name: Representation Transfer: Complexification, Section 7.3
Scenario: Complexification is applicable when the parameterization of a function approximator can change over time. **When to use:** If a function approximator for a task can change its representation over time so that it better represents the true value function, and the function approximator exhibits locality (such as in a CMAC or RBF), Complexification may significantly improve learning performance.
Method Name: Offline Representation Transfer, Section 7.3
Scenario: Offline Representation Transfer is applicable when the representation changes in a task, or for inter-task transfer. **When to use:** Offline Representation transfer is very flexible method, but less effective than other methods – it should be used when other methods are inapplicable.
Method Name: Learning Mappings via Classification, Section 8.1
Scenario: If appropriate inter-task mappings are unknown, but the agent is provided per-object state variable groupings, or the source task and target task have the same state variables, this mapping-learning method is applicable. **When to use:** Experiments show that learning a mapping to enable transfer is significantly better than not using transfer, or using an uninformed mapping that averages together all state variables and actions. If the agent has domain knowledge specifying how state variables can be assigned to objects, but the inter-task mappings are unknown, this method should be used.
Method Name: MASTER, Section 8.2
Scenario: Inter-task mappings are unknown. **When to use:** If the state variables and/or actions in a pair of tasks are different, and Learning Mappings via Classification is not applicable, MASTER should be used to learn appropriate inter-task mappings.

different base RL methods, and with different function approximators. Furthermore, we have shown that these mappings can be learned autonomously using off-the-shelf machine learning algorithms and data gathered by agents in pairs of tasks.

9.2 Possible Enhancements to Monograph Methods

In this section of the monograph we present a number of possible enhancements for the methods introduced by this monograph.

9.2.1 Inter-Task Mappings

Inter-task mappings have been defined so that the state variables and actions are mapped independently. This formulation was sufficient for all the source task and target task pairs considered in this work. However, there may be tasks where these two mappings are interdependent. For instance, it could be that the actions map differently depending on the agent's current location in state space. A second possibility would be tasks in which a mapping between states, rather than state variables, was more effective for transfer. If there are pairs of tasks which exhibit either of these conditions, the inter-task mapping construct would need to be enhanced to account for such possibilities.

None of the transfer methods discussed is able to explicitly take advantage of any knowledge about changes in the reward function between tasks. For example, if it was known that the target task rewards were twice that of the source task, it is possible that value-function methods may be able to automatically modify the source task value function with this background knowledge to enhance learning. As a second example, consider a pair of tasks where the goal state was moved from one edge of the state space to the opposite edge. While the learned transition information could be reused, the policy or value-function would need to be significantly altered to account for the new reward function. It is possible that inter-task mappings could be extended to account for changes in R between tasks, in addition to changes in A and to the set of state variables.

9.2.2 *Transfer Algorithms*

This section enumerates some of the ways in which transfer algorithms presented in this monograph may be improved or expanded upon.

9.2.2.1 Transfer Functionals

Value Function Transfer and Policy Transfer introduced a number of different transfer functionals, denoted ρ. Although all transfer functionals relied on the same inter-task mappings, different ρs were needed for dissimilar function approximators. Given the similarity inherent in the different ρ formulations, it may be possible to construct a transfer functional general enough to work with all of the function approximators considered in this monograph. Ideally, the transfer functional would be function-approximator independent: it would take as inputs an inter-task mapping, a source action-value function, and some details about the function approximator, and then output an appropriate target task action-value function.

When discussing ρ_{ANN}, we posited that "if the target task ANN had additional hidden nodes, a more sophisticated δ mapping [that is used to map hidden nodes from one network to another] could be utilized," as the current transfer δ function assumes that the source and target neural networks had the same number of hidden nodes. If there were more hidden nodes in the target network than in the source network, ρ could distribute the value of weights of a single source task hidden node across multiple target task hidden nodes. Another option would be to connect the additional target task nodes with zero weights so that they would have no impact in initial calculations of the value function, but could be learned over time. If there were fewer hidden nodes in the target network than in the source network, weights from multiple source task hidden nodes could be combined to weight connections to a single target task hidden node, but such a transfer may lose a significant amount of useful information.

9.2.2.2 Increasing Methods' Applicability

TIMBREL was developed to work with instance-based model-learning methods but was only tested using Fitted R-MAX as a base RL method. We predict that it would also work, possibly with minor modifications, in conjunction with other model-based RL algorithms. For instance, TIMBREL could be directly used in the planning phase of Dyna-Q as a source of simulated experience when the agent's model is poor (such as at the beginning of learning a target task). TIMBREL should also be useful, without modification, in R-MAX. In the future we would like to experiment with other model-based RL algorithms to empirically compare the effectiveness of transfer with Fitted R-MAX. Additionally, we intend to apply TIMBREL to more complex domains that have continuous state variables; we expect that transfer will provide even more benefit as task difficulty increases.

The transfer methods in this monograph have modified source task knowledge, whether instances, value functions, policies, or rules, using inter-task mappings. However, our methods do not explicitly account for scaling differences. For instance, recall that Rule Transfer relies on the `Translate()` function to modify a learned source decision list. Rules can be in the form "IF $dist(K_1, T_1) < 5$ THEN Pass to K_3." But if the source task state variables represented distance in meters and the target task represented distance in inches, the constant in this rule would need to be scaled. Similarly, if the source task reward for 3 vs. 2 Keepaway was +1 at every timestep, but the target task reward for 4 vs. 3 Keepaway was changed to +10 at every timestep, the Value Function Transfer functional should be augmented to scale the transferred Q-values appropriately. In both cases we currently rely on a human to scale the rules or Q-values if necessary, but ideally such scaling factors could be learned autonomously, possibly as part of a mapping-learning algorithm.

When discussing Rule Transfer we emphasized that the source task learner was unconstrained but that the target task learned had to use TD learning. However, this requirement is only due to the three rule utilization methods, all of which were designed to affect how the target task agent learns an action-value function. It may be possible to modify one or more of rule utilization methods to allow for other types of base learners in the target task, which would greatly increase the method's applicability.

The Representation Transfer methods, rather than transferring between tasks, were designed to transfer between different representations. However, they can also be combined with inter-task mappings to transfer between tasks as well (as seen in Section 7.3.6). It would also be useful to test if transfer was feasible (and beneficial) if *both* the representation and the task changed.

9.2.2.3 Transfer from Complex to Simple Tasks

This monograph, and the majority of existing TL work, focuses on transfer from a relatively simple source task to a more complex target task. In principle, transferring from complex to simple tasks should also be effective. We do not explore such an option because it would likely only be beneficial in the target task training time scenario (and would not reduce the total training time), but such situations may appear in practice.

9.2.2.4 Learning Partial Mappings

Both mapping learning methods presented in this monograph assume that each action or state variable in the target task has some corresponding action or state variable in the source task. However, for some pairs of tasks, a partial inter-task mapping may be optimal because of some novel target task characteristic which has no correspondence in the source task. For instance, when transferring from the Standard 2D Mountain Car task to the 3D Hand Brake Mountain Car task, the hand-brake action would likely have no correlation with any source task action. This situation could be addressed in MASTER by setting a maximum acceptable MSE threshold, above which actions or state variables could be assigned "cannot map" values, but it may be possible to construct a more robust mechanism.

9.2.2.5 Background Domain Knowledge

Mapping Learning via Classification relies on domain knowledge that groups state variables with task objects. This type of domain knowledge was chosen primarily to be consistent with previous work in mapping learning [Soni and Singh (2006), Talvitie and Singh (2007)]. However, this suggests that there may be other types of semantic knowledge that could also assist with mapping learning. In the future, we would like to explore other types of knowledge that could be provided by a human, or learned, to empower mapping-learning methods without resorting to searching through the full space of possible mappings, as done in MASTER.

9.2.2.6 Reducing Computational Complexity

The primary motivation for MASTER is to learn mappings with few samples at the expense of high computational complexity. We do so under the assumption that for many fielded agents, sample complexity is much more of a bottleneck than computational complexity. In the future we would also like to reduce the computational complexity.

The first area for improvement would be tackling the inner loop of MASTER which is exponential in the number of state variables and actions. As suggested in Section 8.2, to scale this approach to tasks with hundreds of state variables or actions, some type of heuristic search could be included in the algorithm.

Lastly, MASTER relies on agents' ability to explore the target task quickly and build an approximate model. However, in some tasks the initial exploration may not be indicative of the entire MDP, or the target task may be too complex to model, and a model learned with only a little training data would be misleading. While there may be no way to guard against this for arbitrary MDPs, it would be useful to define the type of task for which such initial exploration is likely to yield a useful model for learning a mapping.

9.3 Determining the Efficacy of Transfer

Another question no work in this manuscript directly addresses is how to determine the optimal amount of source task training to minimize the target task training time or total training time. If the source task and target task were identical, the goal of reducing the target task training time would be trivial (by maximizing the source task training time) and the goal of minimizing total training time would be impossible. On the other hand, if the source task and target task were unrelated, it would be impossible to reduce the target task training time through transfer and the total training time would be minimized by not training in the source task at all. It is likely that a calculation or heuristic for determining the optimal amount of source task training time will have to consider the structure of the two tasks, their relationship, and what transfer method is used. This optimization becomes even more difficult in the case of multi-step transfer, as there are two or more tasks that can be trained for different amounts of time.

More fundamental than the amount of time to learn in a source task is the question of whether two tasks are similar enough that transfer may be beneficial. As discussed in Section 5.4, we hypothesize that the main requirement for Value Function Transfer or Q-Value Reuse to successfully improve target task learning is that, on average, at least one of the following is true:

1. The best learned actions in the source task, for a given state, are mapped to the best action in the target task via the inter-task mappings.
2. The average Q-values learned for states are of the correct magnitude in the target task.

Policy Transfer and Rule Transfer are concerned with action selection, not transfer of Q-values, and thus need only be concerned with the first condition above. Instance-based transfer (as done in TIMBREL) uses source task instances to help construct a target task model; the primary requirement is that the model generated in the target task from the source task instances improves planning in the target task, relative to having no source task data. However, as discussed below, none of these guidelines can guarantee that transfer will be productive, or even that transfer will not harm the learner's performance.

9.3.1 Avoiding Negative Transfer

The majority of TL work in the literature has concentrated on showing that a particular transfer approach is plausible. None, to our knowledge, has a well-defined method for determining *when* an approach will fail according to one or more metrics. While we can say that it is possible to improve learning in a target task via transfer, we cannot currently decide if an arbitrary pair of tasks are appropriate for a given transfer method. Therefore, transfer may produce incorrect learning biases and result in negative transfer.

Methods such as MASTER [Taylor et al. (2008d)], which can measure task similarity via model prediction error, or region transfer [Lazaric (2008)], which examines the similarity of tasks at a local level rather than at a per-task level, can help assist when deciding if the agent should transfer or what the agent should transfer. However, neither method provides any theoretical guarantees about its effectiveness.

Consider the source task in Figure 9.1 (top), which is deterministic and discrete. The agent begins in state I and has one action available: `East`. Other states in the "hallway" have two applicable actions: `East` and `West`, except for state A, which also has the actions `North` and `South`. Once the agent executes `North` or `South` in state A, it will remain in state B or C (respectively) and continue self-transitioning. No transition has a reward, except for the self-transition in state B.

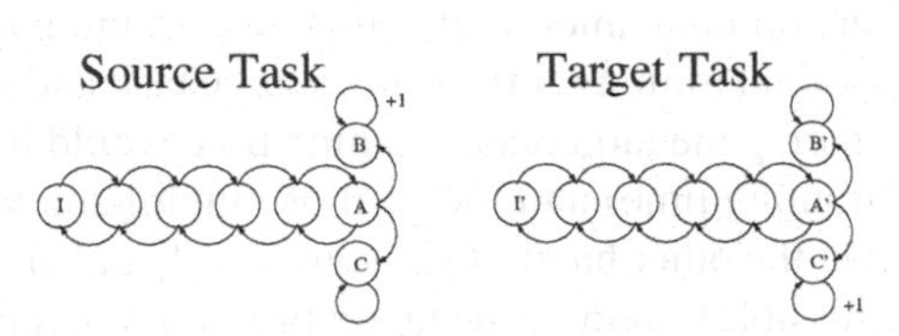

Fig. 9.1 An example pair of tasks that is likely to result in negative transfer for TL methods

Now consider the target task in Figure 9.1 (bottom), which is the same as the source task, except that now the self-transition from C' is the only rewarded transition in the MDP. $Q^\star$(I',`East`) in the target task (the optimal action-value function, evaluated at the state I') is the same as $Q^\star$(I, `East`) in the source task. Indeed, the optimal policy in the target task differs at only a single state, A', and the optimal action-value functions differ only at states A', B', and C'.

One potential method for avoiding negative transfer is to leverage the ideas of *bisimulation* [Milner (1982)]. [Ferns et al. (2006)] point out that:

> In the context of MDPs, bisimulation can roughly be described as the largest equivalence relation on the state space of an MDP that relates two states precisely when for every action, they achieve the same immediate reward and have the same probability of transitioning to classes of equivalent states. This means that bisimilar states lead to essentially the same long-term behavior.

However, bisimulation may be too strict because states are either equivalent or not, and may be slow to compute in practice. The work of [Ferns et al. (2005), Ferns et al. (2006)] relaxes the idea of bisimulation to that of a (pseudo)metric that can be computed much faster, and gives a similarity measure, rather than a boolean. It is possible, although not yet shown, that bisimulation approximations can be used to discover regions of state space that can be transferred from one task to another, or to determine how similar two tasks are *in toto*. In addition to this, or perhaps because of it, there are currently no methods for automatically *constructing* a source task given a target task.[1]

Homomorphisms [Ravindran and Barto (2002)] are a different abstraction that can define transformations between MDPs based on transition and reward dynamics, similar in spirit to inter-task mappings, and have been used successfully

[1] We distinguish this idea from Sherstov and Stone's 2005 approach. Their paper shows it is possible to construct source task perturbations and then allow an agent to spend time learning the set of tasks to attempt to improve learning on an (unknown) source task. Instead, it may be more effective to tailor a source task to a specific target task, effectively enabling an agent to reduce the total number of environmental interactions needed to learn.

for transfer [Soni and Singh (2006)]. However, discovering homomorphisms is NP-hard [Ravindran and Barto (2003a)] and homomorphisms are generally supplied to a learner by an oracle. While these two theoretical frameworks may be able to help avoid negative transfer, or determine when two tasks are "transfer compatible," significant work needs to be done to determine if such approaches are feasible in practice, particularly if the agent is fully autonomous (i.e., is not provided domain knowledge by a human) and is not provided a full model of the MDP.

9.4 Future Transfer Work

As suggested above, TL in RL domains is one area of machine learning where the empirical work has outpaced the theoretical. There is considerable room, and need for, more work that provides bounds, convergence guarantees, and theoretical guarantees of correctness (c.f., [Bowling and Veloso (1999)]). The remainder of this section suggests other open areas.

9.4.1 Increasing Transfer's Applicability

One apparent gap in our taxonomy (Chapter 3) is a dearth of model-learning methods. Because model-learning algorithms are often more sample efficient than model-free algorithms, it is likely that TL will have a large impact on sample complexity when coupled with such efficient RL methods. Moreover, when a full model of the environment is learned in a source task, it may be possible for the target task learner to explicitly reason about how to refine or extend the model as it encounters disparities between it and the target task.

As mentioned in Section 3.3, transfer is an appealing way to set priors in a Bayesian setting. When in a MTL setting, it may be possible to accurately learn priors over a distribution of tasks, enabling a learner to better avoid negative transfer. While there will likely be difficulties associated with scaling up current methods to handle complex tasks, possibly with a complex distribution hierarchy, it seems like Bayesian methods are particularly appropriate for transfer.

Concept drift [Widmer and Kubat (1996)] in RL has not been directly addressed by any work in this monograph. The idea of concept drift is related to a non-stationary environment: at certain points in time, the environment may change arbitrarily. As [Ramon et al. (2007)] note, "for transfer learning, it is usually known when the context change takes place. For concept drift, this change is usually unannounced." Current on-line learning methods may be capable of handling such changes by continually learning. However, it is likely that RL methods developed specifically to converge to a policy and then re-start learning when the concept changes will achieve higher performance, whether such drift is announced or unannounced.

Lastly, none of the methods introduced in this monograph, and few in the related work section, directly address continuing (non-episodic) RL tasks. While TL techniques should directly generalize from the episodic case, it would be useful to verify this claim.

9.4.2 Improving Transfer's Efficacy

The idea of automatically modifying source tasks (c.f., RTP [Sherstov and Stone (2005)], and suggested by [Kuhlmann and Stone (2007)]) has not yet been widely adopted. However, such methods have the potential to improving transfer efficacy in settings where the target task learning performance is paramount. By developing methods that allow training on a sequence of automatically generated variations, TL agents may be able to train autonomously and gain experience that is exploitable in a novel task. Such an approach would be particularly relevant in the multi-task learning setting where the agent could leverage some assumptions about the distribution of the target task(s) it will see in the future.

Transfer provides two distinct benefits to agents in a target task. First, transfer may help improve the agent's exploration so that it discovers higher-valued states more quickly. Secondly, transfer can help bias the agent's internal representation (e.g., its function approximator) so that it may learn faster. It will be important for future work to better distinguish between these two effects; decoupling the two contributions should allow for a better understanding of TL's benefits, as well as provide avenues for future improvements.

Another question not addressed is how to best explore in a source task if the explicit purpose of the agent is to speed up learning in a target task. One could imagine that a non-standard learning or exploration strategy may produce better transfer results, relative to standard strategies. For instance, it may be better to explore more of the source task's state space than to learn an accurate action-value function for only part of the state space. While no current TL algorithms take such an approach, there has been some work on the question of learning a policy that is exploitable (without attempt to maximize the on-line reward accrued while learning) in non-transfer contexts [Şimşek and Barto (2006)].

Similarly, instead of always transferring information from the end of learning in the source task, an agent that knows its information will be used in a target task may decide to record information to transfer partway through training in the source task. For instance [Taylor et al. (2007b)] showed that transfer may be more effective when using policies trained for less time in the source task than when using those trained for more time. Although others have also observed similar behavior [Mihalkova and Mooney (2008)], the majority of work shows that increased performance in the source task is correlated with increased target task performance. For instance, Table 5.3 in Chapter 5 shows that target task agents learn faster with increased amounts of source task training, even after long periods of training in the source task. Understanding how and why this effect occurs will help determine the most appropriate time to transfer information from one task to another.

Given a successful application of transfer, there are potentially two distinct benefits for the agent. First, transfer may help improve the agent's exploration so that it discovers higher-valued states more quickly. Secondly, transfer can help bias the agent's internal representation (e.g., its function approximator) so that it may learn faster. It will be important for future work to better distinguish between these two effects; decoupling the two contributions should allow for a better understanding of TL's benefits, as well as provide avenues for future improvements.

Ideas from *theory revision* [Ginsberg (1988)] (also *theory refinement*) may help inform the automatic construction of inter-task mappings. For example, many methods initialize a target task agent to have Q-values similar to those in the source task agent. Transfer is likely to be successful [Taylor et al. (2007a)] if the target task Q-values are close enough to the optimal Q-values that learning is improved, relative to not using transfer. There are also situations where a *syntactic* change to the knowledge would produce better transfer. For instance, if the target task's reward function were the inverse of the source task function, direct transfer of Q-values would be far from optimal. However, a TL algorithm that could recognize the inverse relationship may be able to use the source task knowledge more appropriately (such as initializing its behavior so that $\pi_{target}(s_{target}) \neq \pi_{source}(\chi_X(s_{target}))$.

Five of the 34 transfer methods in the related work of Chapter 3 [Tanaka and Yamamura (2003), Sunmola and Wyatt (2006), Ferguson and Mahadevan (2006), Lazaric (2008), Wilson et al. (2007)] attempt to discover internal learning parameters (e.g., appropriate features or learning rate) so that future tasks in the same domain may be learned more efficiently. It is likely that other "meta-learning" methods could be useful. For instance, it may be possible to learn to use an appropriate function approximator, an advantageous learning rate, or even the most appropriate RL method. Although likely easier to accomplish in a MTL setting, such meta-learning may also be possible in transfer, given sufficiently strong assumptions about task similarity. Multiple heuristics regarding the best way to select RL methods and learning parameter settings for a particular domain exist, but typically such settings are chosen in an ad hoc manner. Transfer may be able to assist when setting such parameters, rather than relying on human intuition.

Lastly, we hope that the appealing idea of task-invariant knowledge will be extended. Rather than learning an appropriate representation across different tasks, agent-space [Konidaris and Barto (2007)] and RRL techniques attempt to discover knowledge about the agent or the agent's actions which can be directly reused in novel tasks. The better techniques can successfully compartmentalize knowledge, separating what will usefully transfer and what will not, the easier it will be to achieve successful transfer without having to un-learn irrelevant biases.

We now present three possibilities for extending the current RL transfer work to different learning settings in which transfer has not been successfully applied. First, although two of the papers in the related work chapter of this work have examined extensive games [Banerjee and Stone (2007), Kuhlmann and Stone (2007)], none consider stochastic games [Shapley (1953)] or repeated normal form games. For instance, one could consider learning how to play against a set of opponents so that when a new opponent is introduced, the learner may quickly adapt one of its previous strategies rather than completely re-learning a strategy. Another option would be for an agent to learn how to play one game and then transfer the knowledge to a different stochastic game. Due to similarities between RL and these two game playing settings, transfer methods described in this manuscript may be applied with relatively little modification.

A second possibility for extending transfer is into the realm of partially observable MDPs (POMDPs). It may possible to learn a source POMDP and then use knowledge gained to heuristically speed up planning in a target POMDP. Additionally, because it is typically assumed that POMDP planners are given a complete and accurate model of a task, it may be possible to analytically compare source and target tasks before learning in order to determine if transfer would be beneficial, and if so, how best to use the past knowledge.

Third, multi-agent MDP and POMDP learners may also be able to successfully exploit transfer. None of the work in this monograph focuses on multi-agent learning, but it is likely existing methods may be extended to the cooperative multi-agent setting. For instance, when formulating a problem as an MMDP or DEC-MDP, agents must either reason over a joint action space or explicitly reason about how their actions affect others. It may be possible for agents to learn over a subset of actions first, and then gradually add actions (or joint actions) over time, similar to transferring between tasks with different action sets. The need for such speedups is particularly critical in distributed POMDPs, as solving them optimally as been shown to be NEXP-Complete [Bernstein et al. (2002)]. Transfer is one possible approach to making such problems more tractable, but to our knowledge, no such methods have yet been proposed.

Lastly, in order to better evaluate TL methods, it would be helpful to have a standard set of domains and metrics. While classification and regression have long benefited from standard metrics, such as precision and recall, and standard test sets, such as the UC Irvine data sets, there are currently no equivalents for transfer learning. While there is some work in the RL community to standardize on a common interface and set of benchmark tasks [Tanner et al. (2008), Whiteson et al. (2008)], no such standardization has been proposed for the transfer learning in RL community. Even in the absence of such a framework, we suggest that it is important for authors working in this area to:

- Clearly specify the setting: Is the source task learning time discounted? What assumptions are made about the relationship between the source target and target task?
- Evaluate the algorithm with a number of metrics: No one metric captures all possible benefits from transfer.
- Empirically or theoretically compare the performance of novel algorithms: To better evaluate novel algorithms, existing algorithms should be compared using standard metrics on a single task.[2]

As discussed in Section 1.2.1, we do not think that TL for RL methods can be strictly ordered in terms of efficacy, due to the many possible goals of transfer. However,

[2] One of the difficulties inherent in this proposal is that small variations in domain implementation may result in very different learning performances. While machine learning practitioners are able to report past results verbatim when using the same data set, many RL domains used in papers are not released. In order to compare with past work, RL researchers must reimplement, tune, and test past algorithms to compare with their algorithm on their domain implementation.

by standardizing on reporting methodology, TL algorithms can be more easily compared, making it easier to select an appropriate method in a given experimental setting.

9.4.3 Transfer in More Difficult Tasks

Lastly, TL research should be applied to more difficult tasks in the future. Although the Keepaway domain used in this monograph is fairly realistic, it is still learnable by off-the-shelf Sarsa. An impressive demonstration of TL would be to enable agents to learn on a task that was otherwise unlearnable, either because learning without correct biases caused the agent to fail to make any headway, or because a real-world problem has constraints on the amount of data gathered. An example of the first is simulated robot soccer – to date, no research has successfully learned behavior for a full team of agents. An example of the second domain is physical agents – by using transfer, RL methods should be able to learn robotic control significantly faster than learning without transfer, potentially making RL a more practical tool for roboticists.

9.5 Conclusion

This monograph has presented inter-task mappings, a set of transfer learning algorithms that utilize these mappings to transfer between tasks with different actions and state variables, and a set of algorithms capable of learning inter-task mappings. Efficacy of these techniques has been empirically studied using a variety of base RL algorithms, in multiple domains, and with varying amounts of domain knowledge. Taken as a whole, this monograph has introduced a powerful tool for improving the learning capabilities of existing RL algorithms by generalizing across tasks, allowing difficult tasks to be learned faster than if transfer were not used, or possibly learning tasks which would otherwise be unlearnable.

In the coming years deployed agents will become increasingly common and gain more capabilities. While current agents are typically designed to perform a single task, agents will soon be expected to perform multiple tasks. Transfer learning is one way to help enable such multi-purpose agents to train in the real world. Transfer learning, paired with RL, is an appropriate paradigm if the agent must take sequential actions in the world and the designers do not know optimal solutions to the agent's tasks (or even if they do not know what the agent's tasks will be). Before such agents could be reliably deployed, transfer methods will need to be improved, as discussed earlier in this chapter. Additionally, base RL methods will also likely need to be made more sample efficient, and better heuristics need to be developed to determine what type of RL algorithm is most appropriate for a given task. I will consider this research a success if, within a decade, transfer has helped to create a multi-task autonomous agent capable of on-line learning in the real world.

Significant progress on transfer for RL domains has been made in the last few years, but there are many open questions. We expect that many of the above questions will be addressed in the near future so that TL algorithms become more powerful and more broadly applicable. Additionally, we hope to see more physical and virtual agents utilizing transfer learning, further encouraging growth in an exciting and active subfield of the AI community.

Appendix A
On-Line Appendix

An on-line appendix to this book may be found at
`http://teamcore.usc.edu/taylorm/Springer09`.
At this webpage is errata for the monograph as well as code for:

Domains introduced in this monograph

1. 3D Mountain Car Task
2. Inaccurate 3 vs. 2 and Inaccurate 4 vs. 3 Keepaway
3. 3 vs. 2 XOR Keepaway
4. Ringworld
5. Knight Joust

Selected TL algorithms introduced in this monograph

1. The ρ functional used by Value Function Transfer
2. Policy Transfer

M.E. Taylor: Transfer in Reinforcement Learning Domains, SCI 216, pp. 219–219.
springerlink.com

References

[Aamodt and Plaza (1994)] Aamodt, A., Plaza, E.: Case-based reasoning: Foundational issues, methodological variations, and system approaches (1994)

[Abbeel and Ng (2005)] Abbeel, P., Ng, A.Y.: Exploration and apprenticeship learning in reinforcement learning. In: ICML 2005: Proceedings of the 22nd International Conference on Machine Learning, pp. 1–8 (2005)

[Ahmadi et al. (2007)] Ahmadi, M., Taylor, M.E., Stone, P.: IFSA: Incremental Feature-Set Augmentation for reinforcement learning tasks. In: The Sixth International Joint Conference on Autonomous Agents and Multiagent Systems (2007)

[Albus (1981)] Albus, J.S.: Brains, Behavior, and Robotics. Byte Books, Peterborough (1981)

[Andre and Russell (2002)] Andre, D., Russell, S.J.: State abstraction for programmable reinforcement learning agents. In: Proc. of the Eighteenth National Conference on Artificial Intelligence, pp. 119–125 (2002)

[Andre and Teller (1999)] Andre, D., Teller, A.: Evolving team Darwin United. In: Asada, M., Kitano, H. (eds.) RoboCup 1998. LNCS, vol. 1604, pp. 346–351. Springer, Heidelberg (1999)

[Argyrious et al. (2007)] Argyrious, A., Evgenion, T., Pontil, M.: Multitask reinforcement learning on the distribution of MDPs. Machine Learning (2007)

[Asada et al. (1994)] Asada, M., Noda, S., Tawaratsumida, S., Hosoda, K.: Vision-based behavior acquisition for a shooting robot by using a reinforcement learning. In: Proceedings of IAPR/IEEE Workshop on Visual Behaviors 1994, pp. 112–118 (1994)

[Asadi and Huber (2007)] Asadi, M., Huber, M.: Effective control knowledge transfer through learning skill and representation hierarchies. In: Proceedings of the 20th International Joint Conference on Artificial Intelligence, pp. 2054–2059 (2007)

[Atkeson and Santamaria (1997)] Atkeson, C.G., Santamaria, J.C.: A comparison of direct and model-based reinforcement learning. In: Proceedings of the 1997 International Conference on Robotics and Automation (1997)

[Banerjee and Stone (2007)] Banerjee, B., Stone, P.: General game learning using knowledge transfer. In: The 20th International Joint Conference on Artificial Intelligence, pp. 672–677 (2007)

[Banerjee et al. (2006)] Banerjee, B., Liu, Y., Youngblood, G.M.: ICML workshop on Structural Knowledge Transfer for Machine Learning (2006)

[Baxter and Bartlett (2001)] Baxter, J., Bartlett, P.L.: Infinite-horizon policy-gradient estimation. Journal of Artificial Intelligence Research 15, 319–350 (2001)

[Beielstein and Markon (2002)] Beielstein, T., Markon, S.: Threshold selection, hypothesis tests and DOE methods. In: Proceedings of the Congresss on Evolutionary Computation, pp. 777–782 (2002)

[Bellman (1956)] Bellman, R.E.: A problem in the sequential design of experiments. Sankhya 16, 221–229 (1956)

[Bellman (1957)] Bellman, R.E.: Dynamic Programming. Princeton University Press, Princeton (1957)

[Bernstein et al. (2002)] Bernstein, D.S., Givan, R., Immerman, N., Zilberstein, S.: The complexity of decentralized control of Markov Decision Processes. Mathematics of Operations Research 27(4), 819–840 (2002)

[Bowling and Veloso (1999)] Bowling, M.H., Veloso, M.M.: Bounding the suboptimality of reusing subproblem. In: IJCAI 1999: Proceedings of the Sixteenth International Joint Conference on Artificial Intelligence, San Francisco, CA, USA, pp. 1340–1347 (1999)

[Boyan and Moore (1995)] Boyan, J.A., Moore, A.W.: Generalization in reinforcement learning: Safely approximating the value function. In: Tesauro, G., Touretzky, D.S., Leen, T.K. (eds.) Advances in Neural Information Processing Systems, vol. 7, pp. 369–376. MIT Press, Cambridge (1995)

[Bradtke and Duff (1995)] Bradtke, S.J., Duff, M.O.: Reinforcement learning methods for continuous-time Markov decision problems. In: Tesauro, G., Touretzky, D., Leen, T. (eds.) Advances in Neural Information Processing Systems, vol. 7, pp. 393–400. Morgan Kaufmann, San Mateo (1995)

[Brafman and Tennenholtz (2002)] Brafman, R.I., Tennenholtz, M.: R-Max – a general polynomial time algorithm for near-optimal reinforcement learning. Journal of Machine Learning Research 3, 213–231 (2002)

[Carroll and Seppi (2005)] Carroll, J.L., Seppi, K.: Task similarity measures for transfer in reinforcement learning task libraries. In: Proceedings of 2005 IEEE International Joint Conference on Neural Networks, vol. 2, pp. 803–808 (2005)

[Caruana (1995)] Caruana, R.: Learning many related tasks at the same time with backpropagation. In: Advances in Neural Information Processing Systems, vol. 7, pp. 657–664 (1995)

[Chen et al. (2003)] Chen, M., Foroughi, E., Heintz, F., Kapetanakis, S., Kostiadis, K., Kummeneje, J., Noda, I., Obst, O., Riley, P., Steffens, T., Wang, Y., Yin, X.: Users manual: RoboCup soccer server manual for soccer server version 7.07 and later (2003), http://sourceforge.net/projects/sserver/

[Choi et al. (2007)] Choi, D., Konik, T., Nejati, N., Park, C., Langley, P.: Structural transfer of cognitive skills. In: Proceedings of the Eighth International Conference on Cognitive Modeling (2007)

[Cohen (1995)] Cohen, W.W.: Fast effective rule induction. In: International Conference on Machine Learning, pp. 115–123 (1995)

[Colombetti and Dorigo (1993)] Colombetti, M., Dorigo, M.: Robot Shaping: Developing Situated Agents through Learning. Tech. Rep. TR-92-040, International Computer Science Institute, Berkeley, CA (1993)

[Crites and Barto (1996)] Crites, R.H., Barto, A.G.: Improving elevator performance using reinforcement learning. In: Touretzky, D.S., Mozer, M.C., Hasselmo, M.E. (eds.) Advances in Neural Information Processing Systems, vol. 8, pp. 1017–1023. MIT Press, Cambridge (1996)

[Croonenborghs et al. (2007)] Croonenborghs, T., Driessens, K., Bruynooghe, M.: Learning relational options for inductive transfer in relational reinforcement learning. In: Proceedings of the Seventeenth Conference on Inductive Logic Programming (2007)

[DARPA (2005)] DARPA, Transfer learning proposer information pamphlet, BAA #05-29 (2005)

[Dean and Givan (1997)] Dean, T., Givan, R.: Model minimization in Markov decision processes. In: Proceedings of the Thirteenth National Conference on Artificial Intelligence, pp. 106–111 (1997)

[Dearden et al. (1998)] Dearden, R., Friedman, N., Russell, S.J.: Bayesian Q-learning. In: Proceedings of the Fourteenth National Conference on Artificial Intelligence, pp. 761–768 (1998)

[Dearden et al. (1999)] Dearden, R., Friedman, N., Andre, D.: Model based Bayesian exploration. In: Proceedings of the 1999 conference on Uncertainty in AI, pp. 150–159 (1999)

[Dempster et al. (1977)] Dempster, A., Laird, N., Rubin, D.: Maximum-likelihood from incomplete data via the EM algorithm. J. Royal Statistical Soc. Set B (methodological) 39, 1–38 (1977)

[Dietterich (2000)] Dietterich, T.G.: Hierarchical reinforcement learning with the MAXQ value function decomposition. Journal of Artificial Intelligence Research 13, 227–303 (2000)

[Drummond (2002)] Drummond, C.: Accelerating reinforcement learning by composing solutions of automatically identified subtasks. Journal of Artificial Intelligence Research 16, 59–104 (2002)

[Dzeroski et al. (2001)] Dzeroski, S., Raedt, L.D., Driessens, K.: Relational reinforcement learning. Machine Learning 43(1/2), 5–52 (2001)

[Erez and Smart (2008)] Erez, T., Smart, W.D.: What does shaping mean for computational reinforcement learning? In: Proceedings of the Seventh IEEE International Conference on Development and Learning, pp. 215–219 (2008)

[Ernst et al. (2005)] Ernst, D., Geurts, P., Wehenkel, L.: Tree-based batch mode reinforcement learning. Journal of Machine Learning Research 6, 503–556 (2005)

[Ferguson and Mahadevan (2006)] Ferguson, K., Mahadevan, S.: Proto-transfer learning in Markov decision processes using spectral methods. In: Proceedings of the ICML 2006 Workshop on Structural Knowledge Transfer for Machine Learning (2006)

[Fern et al. (2004)] Fern, A., Yoon, S., Givan, R.: Approximate policy iteration with a policy language bias. In: Thrun, S., Saul, L., Schölkopf, B. (eds.) Advances in Neural Information Processing Systems, vol. 16. MIT Press, Cambridge (2004)

[Fernandez and Veloso (2006)] Fernandez, F., Veloso, M.: Probabilistic policy reuse in a reinforcement learning agent. In: Proceedings of the 5th International Conference on Autonomous Agents and Multiagent Systems (2006)

[Ferns et al. (2005)] Ferns, N., Panangaden, P., Precup, D.: Metrics for Markov decision processes with infinite state spaces. In: Proceedings of the 2005 Conference on Uncertainty in Artificial Intelligence, pp. 201–208 (2005)

[Ferns et al. (2006)] Ferns, N., Castro, P., Panangaden, P., Precup, D.: Methods for computing state similarity in Markov decision processes. In: Proceedings of the 22nd Conference on Uncertainty in Artificial intelligence, pp. 174–181 (2006)

[Fink (1999)] Fink, E.: Automatic representation changes in problem solving. Tech. Rep. CMU-CS-99-150, Department of Computer Science, Carnegie Mellon University (1999)

[Foster and Dayan (2004)] Foster, D., Dayan, P.: Structure in the space of value functions. Machine Learning 49(1/2), 325–346 (2004)

[Ginsberg (1988)] Ginsberg, A.: Theory revision via prior operationalization. In: Proceedings of the 1988 National Conference on Artificial Intelligence, pp. 590–595 (1988)

[Guestrin et al. (2003)] Guestrin, C., Koller, D., Gearhart, C., Kanodia, N.: Generalizing plans to new environments in relational MDPs. In: International Joint Conference on Artificial Intelligence (IJCAI 2003), Acapulco, Mexico (2003)

[Ilghami et al. (2005)] Ilghami, O., Munoz-Avila, H., Nau, D.S., Aha, D.W.: Learning approximate preconditions for methods in hierarchical plans. In: ICML 2005: Proceedings of the 22nd International Conference on Machine learning, pp. 337–344 (2005)

[Iscen and Erogul (2008)] Iscen, A., Erogul, U.: A new perspective to the keepaway soccer: The takers. In: Proceedings of the Seventh International Joint Conference on Autonomous Agents and Multi–Agent Systems (AAMAS 2008), Estoril, Portugal (2008)

[Jong and Stone (2007)] Jong, N.K., Stone, P.: Model-based exploration in continuous state spaces. In: The Seventh Symposium on Abstraction, Reformulation, and Approximation (2007)

[Kaelbling et al. (1996)] Kaelbling, L.P., Littman, M.L., Moore, A.W.: Reinforcement learning: A survey. Journal of Artificial Intelligence Research 4, 237–285 (1996)

[Kaelbling et al. (1998)] Kaelbling, L.P., Littman, M.L., Cassandra, A.R.: Planning and acting in partially observable stochastic domains. Artificial Intelligence 101(1-2), 99–134 (1998)

[Kalmár and Szepesvári (1999)] Kalmár, Z., Szepesvári, C.: An evaluation criterion for macro learning and some results. Tech. Rep. TR-99-01, Mindmaker Ltd (1999)

[Kaplan (1989)] Kaplan, C.A.: Switch: A simulation of representational change in the mutilated checkerboard problem. Technical Report C.I.P. 477, Department of Psychology, Carnegie Mellon University (1989)

[Kearns and Singh (1998)] Kearns, M., Singh, S.: Near-optimal reinforcement learning in polynomial time. In: Proc. 15th International Conf. on Machine Learning, pp. 260–268. Morgan Kaufmann, San Francisco (1998)

[Kephart and Chess (2003)] Kephart, J.O., Chess, D.M.: The vision of autonomic computing. IEEE Computer, 41–50 (2003)

[Knox and Stone (2008)] Knox, W.B., Stone, P.: TAMER: Training an Agent Manually via Evaluative Reinforcement. In: IEEE 7th International Conference on Development and Learning (2008)

[Kolter et al. (2008)] Kolter, J.Z., Abbeel, P., Ng, A.: Hierarchical apprenticeship learning with application to quadruped locomotion. In: Platt, J., Koller, D., Singer, Y., Roweis, S. (eds.) Advances in Neural Information Processing Systems, vol. 20, pp. 769–776. MIT Press, Cambridge (2008)

[Konidaris and Barto (2006)] Konidaris, G., Barto, A.: Autonomous shaping: Knowledge transfer in reinforcement learning. In: Proceedings of the 23rd International Conference on Machine Learning, pp. 489–496 (2006)

[Konidaris and Barto (2007)] Konidaris, G., Barto, A.G.: Building portable options: Skill transfer in reinforcement learning. In: Proceedings of the 20th International Joint Conference on Artificial Intelligence, pp. 895–900 (2007)

[Kuhlmann and Stone (2007)] Kuhlmann, G., Stone, P.: Graph-based domain mapping for transfer learning in general games. In: Proceedings of The Eighteenth European Conference on Machine Learning (2007)

[Laird et al. (1986)] Laird, J.E., Rosenbloom, P.S., Newell, A.: Chunking in Soar: The anatomy of a general learning mechanism. Machine Learning 1(1), 11–46 (1986)

[Laird et al. (1987)] Laird, J.E., Newell, A., Rosenbloom, P.S.: Soar: an architecture for general intelligence. Artif. Intell. 33(1), 1–64 (1987)

[Lazaric (2008)] Lazaric, A.: Knowledge transfer in reinforcement learning. PhD thesis, Politecnico di Milano (2008)

[Leffler et al. (2007)] Leffler, B.R., Littman, M.L., Edmunds, T.: Efficient reinforcement learning with relocatable action models. In: Proceedings of the 22nd AAAI Conference on Artificial Intelligence, pp. 572–577 (2007)

[Li et al. (2006)] Li, L., Walsh, T.J., Littman, M.L.: Towards a unified theory of state abstraction for MDPs. In: Proceedings of the Ninth International Symposium on Artificial Intelligence and Mathematics, pp. 531–539 (2006)
[Liu and Stone (2006)] Liu, Y., Stone, P.: Value-function-based transfer for reinforcement learning using structure mapping. In: Proceedings of the Twenty-First National Conference on Artificial Intelligence, pp. 415–420 (2006)
[Maclin and Shavlik (1996)] Maclin, R., Shavlik, J.W.: Creating advice-taking reinforcement learners. Machine Learning 22(1-3), 251–281 (1996)
[Maclin et al. (2005)] Maclin, R., Shavlik, J., Torrey, L., Walker, T., Wild, E.: Giving advice about preferred actions to reinforcement learners via knowledge-based kernel regression. In: Proceedings of the 20th National Conference on Artificial Intelligence (2005)
[Madden and Howley (2004)] Madden, M.G., Howley, T.: Transfer of experience between reinforcement learning environments with progressive difficulty. Artificial Intelligence Review 21(3-4), 375–398 (2004)
[Mahadevan and Connell (1991)] Mahadevan, S., Connell, J.: Automatic programming of behavior-based robots using reinforcement learning. In: National Conference on Artificial Intelligence, pp. 768–773 (1991)
[Mahadevan and Maggioni (2007)] Mahadevan, S., Maggioni, M.: Proto-value functions: A Laplacian framework for learning representation and control in Markov decision processes. Journal of Machine Learning Research 8, 2169–2231 (2007)
[Mataric (1994)] Mataric, M.J.: Reward functions for accelerated learning. In: International Conference on Machine Learning, pp. 181–189 (1994)
[McCarthy (1964)] McCarthy, J.: A tough nut for proof procedures. Technical Report Sail AI Memo 16, Computer Science Department, Stanford University (1964)
[Mehrotra et al. (1997)] Mehrotra, K., Mohan, C.K., Ranka, S.: Elements of artificial neural networks. MIT Press, Cambridge (1997)
[Mehta et al. (2008)] Mehta, N., Natarajan, S., Tadepalli, P., Fern, A.: Transfer in variable-reward hierarchical reinforcement learning. Machine Learning 73(3), 289–312 (2008)
[Mihalkova and Mooney (2008)] Mihalkova, L., Mooney, R.J.: Transfer learning by mapping with minimal target data. In: Proceedings of the AAAI 2008 Workshop on Transfer Learning for Complex Tasks (2008)
[Mihalkova et al. (2007)] Mihalkova, L., Huynh, T., Mooney, R.: Mapping and revising Markov logic networks for transfer learning. In: Proceedings of the 22nd AAAI Conference on Artificial Intelligence (2007)
[Milner (1982)] Milner, R.: A Calculus of Communicating Systems. Springer, New York (1982)
[Moore (1991)] Moore, A.: Variable resolution dynamic programming: Efficiently learning action maps in multivariate real-valued state-spaces. In: Machine Learning: Proceedings of the Eighth International Conference (1991)
[Moore and Atkeson (1993)] Moore, A.W., Atkeson, C.G.: Prioritized sweeping: Reinforcement learning with less data and less real time. Machine Learning 13, 103–130 (1993)
[Ng and Jordan (2000)] Ng, A.Y., Jordan, M.: PEGASUS: A policy search method for large MDPs and POMDPs. In: Proceedings of the 16th Conference on Uncertainty in Artificial Intelligence (2000)
[Ng et al. (1999)] Ng, A.Y., Harada, D., Russell, S.: Policy invariance under reward transformations: Theory and application to reward shaping. In: Proceedings of the 16th International Conference on Machine Learning (1999)
[Ng et al. (2004)] Ng, A.Y., Coates, A., Diel, M., Ganapathi, V., Schulte, J., Tse, B., Berger, E., Liang, E.: Inverted autonomous helicopter flight via reinforcement learning. In: International Symposium on Experimental Robotics (2004)

[Noda et al. (1998)] Noda, I., Matsubara, H., Hiraki, K., Frank, I.: Soccer server: A tool for research on multiagent systems. Applied Artificial Intelligence 12, 233–250 (1998)

[Ormoneit and Sen (2002)] Ormoneit, D., Sen, S.: Kernel-based reinforcement learning. Machine Learning 49(2-3), 161–178 (2002)

[Perkins and Precup (1999)] Perkins, T.J., Precup, D.: Using options for knowledge transfer in reinforcement learning. Tech. Rep. UM-CS-1999-034, The University of Massachusetts at Amherst (1999)

[Potter and De Jong (2000)] Potter, M.A., De Jong, K.A.: Cooperative coevolution: An architecture for evolving coadapted subcomponents. Evolutionary Computation 8, 1–29 (2000)

[Powell (1964)] Powell, M.J.D.: An efficient method for finding the minimum of a function of several variables without calculating derivatives. Computer Journal 7, 155–162 (1964)

[Price and Boutilier (2003)] Price, B., Boutilier, C.: Accelerating reinforcement learning through implicit imitation. Journal of Artificial Intelligence Research 19, 569–629 (2003)

[Puterman (1994)] Puterman, M.L.: Markov Decision Processes: Discrete Stochastic Dynamic Programming. John Wiley & Sons, Inc., Chichester (1994)

[Pyeatt and Howe (2001)] Pyeatt, L.D., Howe, A.E.: Decision tree function approximation in reinforcement learning. In: Proceedings of the Third International Symposium on Adaptive Systems: Evolutionary Computation & Probabilistic Graphical Models, pp. 70–77 (2001)

[Ramon et al. (2007)] Ramon, J., Driessens, K., Croonenborghs, T.: Transfer learning in reinforcement learning problems through partial policy recycling. In: Proceedings of The Eighteenth European Conference on Machine Learning (2007)

[Ravindran and Barto (2002)] Ravindran, B., Barto, A.: Model minimization in hierarchical reinforcement learning. In: Proceedings of the Fifth Symposium on Abstraction, Reformulation and Approximation (2002)

[Ravindran and Barto (2003a)] Ravindran, B., Barto, A.G.: An algebraic approach to abstraction in reinforcement learning. In: Proceedings of the Twelfth Yale Workshop on Adaptive and Learning Systems, pp. 109–114 (2003)

[Ravindran and Barto (2003b)] Ravindran, B., Barto, A.G.: Relativized options: Choosing the right transformation. In: Proceedings of the Twentieth International Conference on Machine Learning (ICML 2003), pp. 608–615. AAAI Press, Menlo Park (2003)

[Riedmiller et al. (2001)] Riedmiller, M., Merke, A., Meier, D., Hoffman, A., Sinner, A., Thate, O., Ehrmann, R.: Karlsruhe brainstormers—a reinforcement learning approach to robotic soccer. In: Stone, P., Balch, T., Kraetzschmar, G.K. (eds.) RoboCup 2000. LNCS, vol. 2019, pp. 367–372. Springer, Heidelberg (2001)

[Roy and Kaelbling (2007)] Roy, D.M., Kaelbling, L.P.: Efficient Bayesian task-level transfer learning. In: Proceedings of the Twentieth International Joint Conference on Artificial Intelligence, Hyderabad, India (2007)

[Rummery and Niranjan (1994)] Rummery, G., Niranjan, M.: On-line Q-learning using connectionist systems. Technical Report CUED/F-INFENG-RT 116, Engineering Department, Cambridge University (1994)

[Saggar et al. (2007)] Saggar, M., D'Silva, T., Kohl, N., Stone, P.: Autonomous learning of stable quadruped locomotion. In: Lakemeyer, G., Sklar, E., Sorrenti, D.G., Takahashi, T. (eds.) RoboCup 2006: Robot Soccer World Cup X. LNCS (LNAI), vol. 4434, pp. 98–109. Springer, Heidelberg (2007)

[Selfridge et al. (1985)] Selfridge, O.G., Sutton, R.S., Barto, A.G.: Training and tracking in robotics. In: Proceedings of the Ninth International Joint Conference on Artificial Intelligence, pp. 670–672 (1985)

[Shapley (1953)] Shapley, L.S.: Stochastic games. Proceedings of the National Academy of Sciences of the United States of America 39(10), 1095–1100 (1953)

[Sharma et al. (2007)] Sharma, M., Holmes, M., Santamaria, J., Irani, A., Isbell, C., Ram, A.: Transfer learning in real-time strategy games using hybrid CBR/RL. In: Proceedings of the Twentieth International Joint Conference on Artificial Intelligence (2007)

[Sherstov and Stone (2005)] Sherstov, A.A., Stone, P.: Improving action selection in MDP's via knowledge transfer. In: Proceedings of the Twentieth National Conference on Artificial Intelligence (2005)

[Silver et al. (2005)] Silver, D., Bakir, G., Bennett, K., Caruana, R., Pontil, M., Russell, S., Tadepalli, P.: NIPS workshop on Inductive Transfer: 10 Years Later (2005)

[Simon (1975)] Simon, H.A.: The functional equivalence of problem solving skills. Cognitive Psychology 7, 268–288 (1975)

[Şimşek and Barto (2006)] Şimşek, Ö., Barto, A.G.: An intrinsic reward mechanism for efficient exploration. In: Proceedings of the Twenty-Third International Conference on Machine Learning (2006)

[Singh and Sutton (1996)] Singh, S., Sutton, R.S.: Reinforcement learning with replacing eligibility traces. Machine Learning 22, 123–158 (1996)

[Singh (1992)] Singh, S.P.: Transfer of learning by composing solutions of elemental sequential tasks. Machine Learning 8, 323–339 (1992)

[Skinner (1953)] Skinner, B.F.: Science and Human Behavior. Colliler-Macmillian (1953)

[Soni and Singh (2006)] Soni, V., Singh, S.: Using homomorphisms to transfer options across continuous reinforcement learning domains. In: Proceedings of the Twenty First National Conference on Artificial Intelligence (2006)

[Srinivasan (2001)] Srinivasan, A.: The Aleph manual (2001)

[Stagge (1998)] Stagge, P.: Averaging efficiently in the presence of noise. In: Eiben, A.E., Bäck, T., Schoenauer, M., Schwefel, H.-P. (eds.) PPSN 1998. LNCS, vol. 1498, pp. 188–200. Springer, Heidelberg (1998)

[Stanley and Miikkulainen (2002)] Stanley, K.O., Miikkulainen, R.: Evolving neural networks through augmenting topologies. Evolutionary Computation 10, 99–127 (2002)

[Stanley and Miikkulainen (2004a)] Stanley, K.O., Miikkulainen, R.: Competitive coevolution through evolutionary complexification. Journal of Artificial Intelligence Research (JAIR) 21, 63–100 (2004)

[Stanley and Miikkulainen (2004b)] Stanley, K.O., Miikkulainen, R.: Evolving a roving eye for go. In: Proceedings of the Genetic and Evolutionary Computation Conference (2004)

[Stone and Veloso (2000)] Stone, P., Veloso, M.: Multiagent systems: A survey from a machine learning perspective. Autonomous Robots 8(3), 345–383 (2000)

[Stone et al. (2005)] Stone, P., Sutton, R.S., Kuhlmann, G.: Reinforcement learning for RoboCup-soccer keepaway. Adaptive Behavior 13(3), 165–188 (2005)

[Sunmola and Wyatt (2006)] Sunmola, F.T., Wyatt, J.L.: Model transfer for Markov decision tasks via parameter matching. In: Proceedings of the 25th Workshop of the UK Planning and Scheduling Special Interest Group (PlanSIG 2006) (2006)

[Sutton (1988)] Sutton, R.S.: Learning to predict by the methods of temporal differences. Machine Learning 3, 9–44 (1988)

[Sutton (1996)] Sutton, R.S.: Generalization in reinforcement learning: Successful examples using sparse coarse coding. In: Advances in Neural Information Processing Systems, vol. 8, pp. 1038–1044 (1996)

[Sutton and Barto (1998)] Sutton, R.S., Barto, A.G.: Introduction to Reinforcement Learning. MIT Press, Cambridge (1998)

[Sutton et al. (1999)] Sutton, R.S., Precup, D., Singh, S.P.: Between MDPs and semi-MDPs: A framework for temporal abstraction in reinforcement learning. Artificial Intelligence 112(1-2), 181–211 (1999)

[Sutton et al. (2007)] Sutton, R.S., Koop, A., Silver, D.: On the role of tracking in stationary environments. In: Proceedings of the 24th International Conference on Machine Learning (2007)

[Swarup and Ray (2006)] Swarup, S., Ray, S.R.: Cross-domain knowledge transfer using structured representations. In: Proceedings of the Twenty First National Conference on Artificial Intelligence (2006)

[Syed and Schapier (2007)] Syed, U., Schapier, R.: A multiplicative weights algorithm for apprenticeship learning. In: Advances in Neural Information Processing Systems, vol. 21 (2007)

[Talvitie and Singh (2007)] Talvitie, E., Singh, S.: An experts algorithm for transfer learning. In: Proceedings of the Twentieth International Joint Conference on Artificial Intelligence (2007)

[Tanaka and Yamamura (2003)] Tanaka, F., Yamamura, M.: Multitask reinforcement learning on the distribution of MDPs. Transactions of the Institute of Electrical Engineers of Japan C 123(5), 1004–1011 (2003)

[Tanner et al. (2008)] Tanner, B., White, A., Sutton, R.S.: RL Glue and Codecs (2008), `http://mloss.org/software/view/151/`

[Taylor and Stone (2004)] Taylor, M.E., Stone, P.: Speeding up reinforcement learning with behavior transfer. In: AAAI 2004 Fall Symposium on Real-life Reinforcement Learning (2004)

[Taylor and Stone (2007a)] Taylor, M.E., Stone, P.: Cross-domain transfer for reinforcement learning. In: Proceedings of the Twenty-Fourth International Conference on Machine Learning (2007)

[Taylor and Stone (2007b)] Taylor, M.E., Stone, P.: Representation transfer for reinforcement learning. In: AAAI 2007 Fall Symposium on Computational Approaches to Representation Change during Learning and Development (2007)

[Taylor et al. (2005)] Taylor, M.E., Stone, P., Liu, Y.: Value functions for RL-based behavior transfer: A comparative study. In: Proceedings of the Twentieth National Conference on Artificial Intelligence (2005)

[Taylor et al. (2006)] Taylor, M.E., Whiteson, S., Stone, P.: Comparing evolutionary and temporal difference methods in a reinforcement learning domain. In: Proceedings of the Genetic and Evolutionary Computation Conference, pp. 1321–1328 (2006)

[Taylor et al. (2007a)] Taylor, M.E., Stone, P., Liu, Y.: Transfer learning via inter-task mappings for temporal difference learning. Journal of Machine Learning Research 8(1), 2125–2167 (2007)

[Taylor et al. (2007b)] Taylor, M.E., Whiteson, S., Stone, P.: Transfer via inter-task mappings in policy search reinforcement learning. In: The Sixth International Joint Conference on Autonomous Agents and Multiagent Systems (2007)

[Taylor et al. (2008a)] Taylor, M.E., Fern, A., Driessens, K., Stone, P., Maclin, R., Shavlik, J.: AAAI workshop on Transfer Learning for Complex Tasks (2008)

[Taylor et al. (2008b)] Taylor, M.E., Jong, N., Stone, P.: Transferring instances for model-based reinforcement learning. In: Proceedings of the Adaptive Learning Agents and Multi-Agent Systems (ALAMAS+ALAG) workshop at AAMAS 2008 (2008)

[Taylor et al. (2008c)] Taylor, M.E., Jong, N.K., Stone, P.: Transferring instances for model-based reinforcement learning. In: Proceedings of the European Conference on Machine Learning and Principles and Practice of Knowledge Discovery in Databases (ECML PKDD), pp. 488–505 (2008)

[Taylor et al. (2008d)] Taylor, M.E., Kuhlmann, G., Stone, P.: Autonomous transfer for reinforcement learning. In: The Seventh International Joint Conference on Autonomous Agents and Multiagent Systems (2008)

[Tesauro (1994)] Tesauro, G.: TD-Gammon, a self-teaching backgammon program, achieves master-level play. Neural Computation 6(2), 215–219 (1994)

[Thorndike and Woodworth (1901)] Thorndike, E., Woodworth, R.: The influence of improvement in one mental function upon the efficiency of other functions. Psychological Review 8, 247–261 (1901)

[Thrun (1996)] Thrun, S.: Is learning the n-th thing any easier than learning the first? In: Advances in Neural Information Processing Systems, vol. 8, pp. 640–646 (1996)

[Torrey et al. (2005)] Torrey, L., Walker, T., Shavlik, J.W., Maclin, R.: Using advice to transfer knowledge acquired in one reinforcement learning task to another. In: Proceedings of the Sixteenth European Conference on Machine Learning, pp. 412–424 (2005)

[Torrey et al. (2006)] Torrey, L., Shavlik, J.W., Walker, T., Maclin, R.: Skill acquisition via transfer learning and advice taking. In: Proceedings of the Sixteenth European Conference on Machine Learning, pp. 425–436 (2006)

[Torrey et al. (2007)] Torrey, L., Shavlik, J.W., Walker, T., Maclin, R.: Relational macros for transfer in reinforcement learning. In: Proceedings of the Seventeenth Conference on Inductive Logic Programming (2007)

[Walsh et al. (2006)] Walsh, T.J., Li, L., Littman, M.L.: Transferring state abstractions between MDPs. In: Proceedings of the ICML 2006 Workshop on Structural Knowledge Transfer for Machine Learning (2006)

[Walsh et al. (2004)] Walsh, W.E., Tesauro, G., Kephart, J.O., Das, R.: Utility functions in autonomic systems. In: Proc. of the International Conf. on Autonomic Computing, pp. 70–77 (2004)

[Watkins (1989)] Watkins, C.J.C.H.: Learning from delayed rewards. PhD thesis, King's College, Cambridge, UK (1989)

[Whiteson and Stone (2006)] Whiteson, S., Stone, P.: Evolutionary function approximation for reinforcement learning. Journal of Machine Learning Research (2006)

[Whiteson et al. (2008)] Whiteson, S., White, A., Tanner, B., Sutton, R.S.S., Precup, D., Stone, P., Littman, M., Vlassis, N., Riedmiller, M.: ICML workshop on The 2008 RL-Competition (2008)

[Widmer and Kubat (1996)] Widmer, G., Kubat, M.: Learning in the presence of concept drift and hidden contexts. Machine Learning 23(1), 69–101 (1996)

[Williams and Singh (1999)] Williams, J.K., Singh, S.: Experimental results on learning stochastic memoryless policies for partially observable Markov decision processes. In: Proceedings of the 1998 conference on Advances in neural information processing systems II, pp. 1073–1079. MIT Press, Cambridge (1999)

[Williams (1992)] Williams, R.J.: Simple statistical gradient-following algorithms for connectionist reinforcement learning. Machine Learning 8, 229–256 (1992)

[Wilson et al. (2007)] Wilson, A., Fern, A., Ray, S., Tadepalli, P.: Multi-task reinforcement learning: a hierarchical Bayesian approach. In: ICML 2007: Proceedings of the 24th international conference on Machine learning, pp. 1015–1022 (2007)

[Winograd (1972)] Winograd, T.: Understanding Natural Language. Edinburgh University Press (1972)

[Witten and Frank (2005)] Witten, I.H., Frank, E.: Data Mining: Practical machine learning tools and techniques. Morgan Kaufmann, San Francisco (2005)

[Yao (1999)] Yao, X.: Evolving artificial neural networks. Proceedings of the IEEE 87(9), 1423–1447 (1999)

[Zhang and Dietterich (1995)] Zhang, W., Dietterich, T.G.: A reinforcement learning approach to job-shop scheduling. In: Proceedings of the International Joint Conference on Artificial Intelligence (1995)